T0321191

Cortical Sensory Organization

Multiple Auditory Areas

Cortical Sensory Organization
Edited by *Clinton N. Woolsey*

Cortical Sensory Organization

Volume 3

Multiple Auditory Areas

Edited by

Clinton N. Woolsey

University of Wisconsin, Madison, Wisconsin

Humana Press • **Clifton, New Jersey**

Library of Congress Cataloging in Publication Date

Main entry under title:

Cortical sensory organization.

  Includes bibliographies and indexes.
  Contents: v. 1. Multiple somatic areas
— v. 3. Multiple auditory areas.
  1. Cerebral cortex.   2. Senses and sensation.
I. Woolsey, Clinton N.
QP383.C67   599.8′04182   81-81433
ISBN 0-89603-030-X (v. 1)
AACR2

© 1982 The HUMANA Press Inc.
Crescent Manor
P. O. Box 2148
Clifton, N. J. 07015

Printed in the United States of America

Table of Contents

Chapter 1
The Auditory Cortex: *Patterns of Corticocortical*
Projections Related to Physiological Maps in the
Cat *1*
Thomas J. Imig, Richard A. Reale and **John F.**
Brugge

Chapter 2

Auditory Forebrain Organization: *Thalamocortical and Corticothalamic Connections in the Cat* 43

Michael M. Merzenich, Steve A. Colwell and Richard A. Andersen

Chapter 3

Auditory Cortical Areas in Primates 59

John F. Brugge

Chapter 4

Organization of Auditory Connections: *The Primate Auditory Cortex.* 71

Kathleen A. FitzPatrick and Thomas J. Imig

Chapter 5
Polysensory "Association" Areas of the Cerebral Cortex: *Organization of Acoustic Input in the Cat* *111*
D. R. F. Irvine and D. P. Phillips

Chapter 8
Cortical Auditory Area of Macaca mulatta and Its Relation to the Second Somatic Sensory Area (SM II): *Determination by Electrical Excitation of Auditory Nerve Fibers in the Spiral Osseous Lamina and by Click Stimulation* 231
C. N. Woolsey and **E. M. Walzl**

Contents of Other Volumes

xi

Preface

In April 1979 a symposium on *"Multiple Somatic Sensory Motor, Visual and Auditory Areas and Their Connectivities"* was held at the FASEB meeting in Dallas, Texas under the auspices of the Committee on the Nervous System of the American Physiological Society. The papers presented at that symposium are the basis of most of the substantially augmented, updated chapters in the three volumes of *Cortical Sensory Organization*. Only material in chapter 8 of volume 3 was not presented at that meeting.

The aim of the symposium was to review the present status of the field of cortical representation in the somatosensory, visual and auditory systems. Since the early 1940s, the number of recognized cortical areas related to each of these systems has been increasing until at present the number of visually related areas exceeds a dozen. Although the number is less for the somatic and auditory systems, these also are more numerous than they were earlier and are likely to increase still further since we may expect each system to have essentially the same number of areas related to it.

Discovery of second somatic, visual and auditory areas in the early 1940s followed soon after the development of the evoked potential method for the study of cortical localization. The great increase in the number of recognized areas in the last 10 years has resulted from the use of microelectrode recordings from small clusters of neurons and from single units, which permit far more detailed examination of the brain than did the technology of earlier years. Other factors have been the study of more lightly anesthetized animals and of unanesthetized animals, whose brains have been explored through chambers implanted over the areas to be studied. One can expect the number of recognized areas to increase as more of the cortical surface is explored in detail in various species of animals.

Most individual studies to date have dealt with a single system for which the investigators have developed specialized equipment and skills in its study. There is evidence, however, that some cortical areas may respond to more than one modality of sensory input. This is particularly true of the so-called "association" cortex of the suprasylvian gyrus of the cat. It now seems very desirable to explore under optimal conditions all cortical areas with stimuli of more than one sensory modality. This will require that investigators acquire sophisticated equipment for the study of somatic, visual and auditory systems and develop skills in its use. An alternative method would be for experts on each system to join forces, so that the methods specialized for the three systems can be applied in a single given experiment.

Increasing quantities of information on the organization of afferent and efferent systems are being derived from the application of techniques for the study of connectivities within the central nervous system, through the use of tritiated amino acids and horseradish peroxidase, as illustrated by several of the studies reported in these volumes.

An important area of research not covered in these volumes is the study of behaving animals with implanted recording electrodes. I foresee that ultimately all areas of the cortex will be examined in this way.

Another area requiring study is the sensory input to the cortical motor areas. Corticocortical connections to these areas have been studied, as reported in volume 1, but more detailed sensory input using electrophysiological methods have not yet defined the sensory inputs to the precentral and supplementary motor areas. Similarly, less work has been done on the motor output from the postcentral sensory areas and its relation to the sensory input to these areas. There is practically no modern work on the effects of electrical stimulation of the visual and auditory areas of the cortex, although motor effects were obtained on stimulation of these areas by Ferrier and other early students of cortical localization.

An important problem for the future concerns the terminology to be applied to the many new cortical areas. If these areas correspond to recognized cytoarchitectural areas of Brodmann, that terminology should be applied. At present there is considerable confusion in terminology, perhaps best illustrated by the terms used to describe the various auditory areas in cat and monkey, where terms for the cat are related to the position of the areas in the auditory region. However, because the auditory region changes its orientation with evolution, the same terms used for cat cannot be used for the monkey. Perhaps when all areas have been identified and their corticortical connections and relations with subcortical structures have been fully defined, a more rational terminology can be proposed.

The three volumes of this work do not include reports from all the important workers in the fields surveyed. It was not possible in the time available to the symposium to include all those we should have liked to invite.

The editor wishes to express his deep appreciation to all those who took part in the Dallas symposium, and to thank them for the manuscripts which they prepared for these three volumes on *Cortical Sensory Organization*. He is also grateful to Drs. J. C. Coulter, J. H. Kaas and J. F. Brugge, who chaired the three programs. Finally, special thanks is due to Thomas Lanigan of the Humana Press for his interest in publishing this work and the care that he has devoted to seeing the material through the press.

The editor also wishes to thank Evadine Olson for several typing tasks that she performed in relation to his editorial functions.

Chapter 1

The Auditory Cortex

Patterns of Corticocortical Projections Related to Physiological Maps in the Cat

Thomas J. Imig,* Richard A. Reale and John F. Brugge

Department of Neurophysiology and Waisman Center on Mental Retardation and Human Development, University of Wisconsin Medical School, Madison, Wisconsin

Auditory cortex receives input from two major fiber systems. One system consists of the thalamocortical fibers that are the final link in the synaptic chain connecting the ear to the cerebral cortex. It is this system that presumably imposes tonotopic and binaural maps upon the auditory cortex. A second input system is composed of fibers that arise and terminate within the cortex itself. These cortico-cortical connections serve to interconnect different areas of the cer-

---

*Current address: Department of Physiology, Kansas University Medical Center, Kansas City, KS 66103.

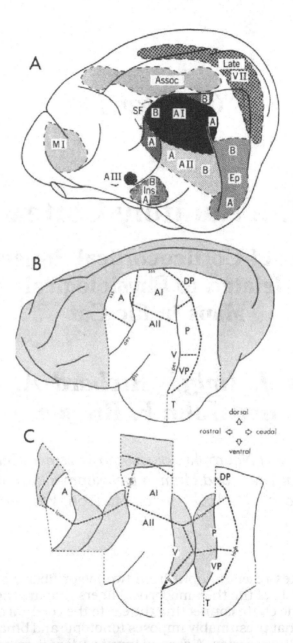

FIG. 1.1. Auditory cortical fields in the cat. A, Woolsey's summary diagram showing four central areas with cochlea represented anteroposteriorly from apex (A) to base (B) in the suprasylvian fringe sector (SF), from
base to apex in A I; from apex to base in A II. In Ep, representation is base
above, apex below. In insula (Ins), evidence suggests base represented
above, apex below. A III is Tunturi's third auditory area. "Association" cortex (Assoc) and precentral motor cortex (M I) gave responses to click with

ebral hemispheres. This report focuses on the relationships of tonotopic and binaural maps to topographic distributions of corticocortical connections within cat auditory cortex.

1. Tonotopic Organization

Several years ago, Woolsey (27, 28) reviewed the available data concerning the tonotopic organization of cat auditory cortex and formulated the map that is pictured in Fig. 1.1A. Within the ectosylvian region, roughly half the auditory cortical surface lies on the exposed gyral surfaces; the remainder is folded into sulci. Based largely on maps of the gyral surfaces, Woolsey divided this auditory region into four tonotopic representations: the first or primary auditory field (A I), the second auditory field (A II), the posterior ectosylvian area (Ep) and the suprasylvian fringe (SF). He also recognized several extraectosylvian auditory response areas, which are labeled M I, Ins, A III, Assoc and Late V II. Recently, a reinterpretation of the tonotopic organization of auditory cortex has been proposed on the basis of studies in which both the sulcal banks and gyral surfaces were mapped with microelectrodes (17, 23). A summary map that takes into account the results of both older evoked potential and more recent microelectrode studies appears in Figs. 1.1B, C. In order to diagram the positions of auditory fields, which extend across gyral surfaces and into sulci, the convolutions of the auditory cortex in the unshaded portion of Fig. 1.1B have been unfolded in Fig. 1.1C to form a flattened two-dimensional surface. Banks of sulci are shaded to distinguish them from gyral surfaces. Light interrupted lines indicate discontinuities on gyral surfaces that were introduced during the unfolding process. Field A I, an anterior auditory field (A), a posterior auditory field (P) and a ventro-

---

FIG. 1.1. (continued)

15-msec latencies under chloralose. Visual area II (Late V II) gave responses with 100-ms latency, also under chloralose (from Woolsey, 28). B, a lateral view of the left cerebral hemisphere showing positions of fields. C, unfolded cortical surface forming the gyral surfaces and sulcal banks in the unshaded region in B; cortical surfaces forming sulcal banks are shaded, while those forming gyral surfaces are not. Heavy interrupted lines delimit four tonotopically organized fields (A, A I, P and VP). The locations of the lowest and highest best-frequencies in these fields are indicated in C by "low" and "high," respectively. Surrounding these four fields is a belt of cortex containing neurons responsive to acoustic stimulation. (B and C from Imig and Reale, 15).

posterior auditory field (VP) are delimited by heavy interrupted lines. Each of these fields contains a complete and orderly tonotopic representation. Surrounding these four fields is a peripheral belt of auditory responsive cortex that includes the second (A II), temporal (T), dorsoposterior (DP) and ventral (V) areas.

1.1. Field A I

The two maps in Fig. 1.1 are largely in agreement with respect to the location and tonotopic organization of A I. The basal or high-frequency end of the cochlea is represented rostrally within the primary field (B in Fig. 1.1A; high in Fig. 1.1C) and the apical or low-frequency end of the cochlea is located caudally (A in Fig. 1.1A; low in Fig. 1.1C). A best-frequency map of A I obtained in one experiment is illustrated in Fig. 1.2. The rostral end of the brain is oriented toward the top of the figure, as indicated in the inset drawing. Within A I, highest best frequencies are located rostrally on the middle ectosylvian gyrus and the lowest best frequencies are located within the dorsal part of the posterior ectosylvian sulcus. Moving across the field in a caudal to rostral direction, ascending sequences of best frequencies are encountered. With respect to these features, the results of recent microelectrode mapping experiments (19, 23) are in accord with earlier evoked potential studies (28, 29).

1.2. Field A

Knight (17) described a tonotopic representation located rostral to A I that he termed the anterior auditory field (A in Fig. 1.1B). Within this field, lowest best frequencies are represented rostroventrally on the anterior ectosylvian gyrus and adjoining sulcal banks, whereas the highest best frequencies adjoin the high best-frequency representation of A I. In Woolsey's summary map, however, a somewhat different pattern of organization was proposed. Woolsey and Walzl (29) described a projection from the apical (low-frequency) end of the cochlea upon the anterior ectosylvian gyrus. Later, a high-frequency focus was described dorsal to the low-frequency focus in A I (4, 11, 12). In his 1960 paper, Woolsey (27) deduced from these data that a "complete frequency representation probably exists along the suprasylvian border of A I, in which the order of frequency representation is the reverse of that in A I" (p. 176). Woolsey and his colleagues tested this hypothesis experimentally and published their results in the following year (28). Their map is reproduced in Fig. 1.3. In response to tones of 0.7 kHz, foci of activity are seen located near the dorsal end of the posterior ecto-

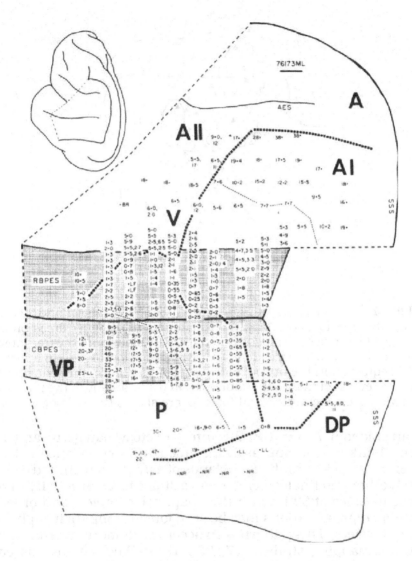

FIG. 1.2. Best-frequency map of auditory cortex obtained in the left hemisphere of one brain. Interrupted lines in the inset brain drawing show the area from which the best-frequency map was constructed. The cortical convolutions in this area were unfolded to produce a flattened two-dimensional surface. The bar below the animal number (76173ML) represents 1 mm on the surface map. Cortical surfaces forming sulcal banks are shaded, while those forming gyral surfaces are not. Heavy interrupted lines indicate physiologically determined field boundaries. Best frequencies are expressed in kHz. Large decimal points are located at recording sites, which were projected onto the cortical surface. Light interrupted lines crossing the cortical surface indicate isofrequency lines (modified from Reale and Imig, 23).

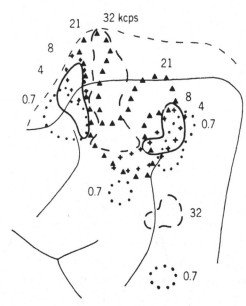

FIG. 1.3. Areas of response to tone pulses of 0.7, 4, 8, 21 and 32 kHz. At upper left is shown overlapping distribution of response areas on lateral bank of suprasylvian sulcus and their extensions onto the free lateral aspect of the cortex. The frequency increases from low to high in a rostro-caudal sequence. In the upper central part of the figure, the sequence in A I is the reverse of that in the suprasylvian fringe sector. In the posterior ectosylvian area, foci for 32 and 0.7 kHz are shown (from Woolsey, 28).

sylvian sulcus(in A I) and on the anterior ectosylvian gyrus and adjoining bank of the suprasylvian sulcus. Tones of successively higher frequencies resulted in islands of activity that shifted closer to each other and finally merged into a single focus of activity in response to tones of 32 kHz. In this map, two complete and orderly tonotopic representations may be seen joined along their high frequency borders. This pattern is in accord with more recent microelectrode mapping studies (17, 23). Thus, Woolsey and his colleagues produced the first and most complete map of the anterior ectosylvian region in the cat and clearly demonstrated its tonotopic organization, although his summary map was not changed to reflect these experimental results. Because the organization of the suprasylvian fringe (SF) as depicted on the summary map may need revision, we use Knight's terminology to designate the anterior auditory field.

1.3. Fields P and VP

Fields P and VP are located in the posterior ectosylvian sulcus (Figs. 1.1B, C) and have only recently been described (23). Field P is located largely on the banks of the posterior ectosylvian sulcus, but extends for a distance upon the posterior ectosylvian gyrus. A nearly complete map of this field obtained in one experiment is seen in Fig. 1.2. The lowest best frequencies in field P adjoin the lowest best frequencies in A I; the highest best frequencies in field P are located ventrally on the posterior ectosylvian gyrus and the superficial portion of the caudal bank of the posterior ectosylvian sulcus. A second tonotopical representation, field VP, lies within the ventral third of the posterior ectosylvian sulcus. This field was incompletely mapped in the experiment illustrated in Fig. 1.2, but a more complete map is illustrated in Fig. 1.4. Fields P and VP share a common border along the mid-to-high best-frequency representations.

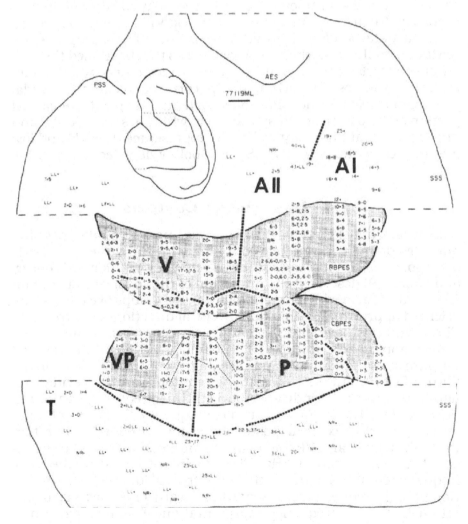

FIG. 1.4. Best-frequency map of auditory cortex in the left hemisphere. See Fig. 1.2 for further details (from Reale and Imig, 23).

The low best-frequency representations of fields P and VP are discontinuous and the lowest best frequencies in VP are located near the ventral end of the posterior ectosylvian sulcus. Although the banks of the posterior ectosylvian sulcus had not been explored when Woolsey formulated his map, the posterior ectosylvian gyrus had been mapped (4, 24, 29). In those evoked potential studies, a representation of the basal (high frequency) end of the cochlea was seen on the posterior ectosylvian gyrus in some maps (e.g., Fig. 1.3 and refs. 4 and 29), which closely corresponds in location to the adjacent high-frequency representations of fields P and VP. Furthermore, Sindberg and Thompson (24) found a representation of the apical (low-frequency) end of the cochlea near the ventral end of the posterior ectosylvian sulcus and gyrus that may correspond in part to the low-frequency representation of field VP. From these evoked potential data, Woolsey deduced the existence of a tonotopic representation on the posterior ectosylvian gyrus that he termed the posterior ectosylvian area, Ep. Although more recent findings (23) suggest that the rostral portion of Ep corresponds to the caudal portions of fields P and VP, evoked potential maps of the caudal portion of the posterior ectosylvian area, which lies outside of P and VP, suggest that high frequencies are represented dorsally and low frequencies ventrally (4, 24, 28), as Woolsey depicted them in his map.

1.4. Isofrequency Contours

Within each of fields A I, A, P and VP neurons with similar best frequencies define isofrequency contours, which are oriented nearly orthogonal to the low-to-high best-frequency gradients. Woolsey and Walzl (29) described the projection of the cochlear nerve upon A I as a "system in which functionally related groups of elements are oriented approximately in a dorso-ventral direction. . . ." (p. 323). These findings have more recently been confirmed in microelectrode mapping experiments (19, 23). In Fig. 1.2 a light interrupted line indicates the course of an isofrequency contour (7.7 kHz) in A I. Within fields P and VP, isofrequency contours are also evident. A light interrupted line indicates the 5.0 kHz isofrequency contour in field P of Fig. 1.2. In Fig. 1.4, two isofrequency contours delimit a curving band of cortical tissue that runs without interruption through fields P and VP and contains neurons with best frequencies between 9 and 13 kHz. Finally, Knight (17) described isofrequency contours in field A that shift from a dorsoventral orientation in high-frequency representation toward a rostrocaudal orientation in the low-frequency representation. A similar pattern of organization can be inferred from Woolsey's map of this region (Fig. 1.3).

1.5. The Peripheral Auditory Belt

Surrounding fields A I, A, P and VP is a belt of cortex that contains neurons responsive to acoustic stimulation. In Figs. 1.1B, C, this belt has been partially subdivided into areas DP, A II, V and T and also includes the caudal part of Woolsey's area Ep and a band of cortex surrounding the portions of field A which do not adjoin A I. DP corresponds to the caudal part of Woolsey's area SF. Although there is some evidence for tonotopy within A II (28), V (23) and the caudal part of Ep (24, 28), the tonotopic organization of the belt has not been worked out in detail.

2. Corticocortical Connections Related to Best-Frequency Maps

2.1. Relationship between Best Frequencies Located at the Sources and Terminations of Corticocortical Connections

In 1960, Downman et al. published the results of experiments in which they studied auditory corticocortical connectivity in the cat using electrical stimulation and evoked potential recording at the cortical surface. In one experiment, they identified two areas on the middle and posterior ectosylvian gyri, from which responses to electrical stimulation of the basal end of the cochlea were recorded. In our maps, these two regions correspond to the high-frequency representations located near the border between fields A I and A and near the border between fields P and VP, respectively. They demonstrated that direct electrical stimulation of the cortical surface in one of these regions evoked responses in the other. From these experiments, they concluded that "some areas of cortex, which are activated by stimulation of a limited portion of the cochlear nerve, are rather specifically interrelated" (p. 139).

Anatomical study of the interconnections between auditory fields (3, 15) has confirmed and extended these findings. During electrophysiological mapping of auditory cortex, tritiated amino acid was injected into the mapped area in order to autoradiographically label the distribution of corticocortical projections. Figure 1.5 illustrates an example of an experiment, in which the isotope was

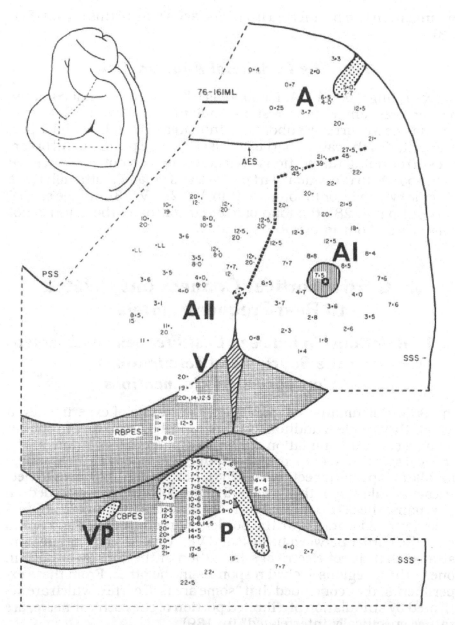

FIG. 1.5. Topographic distribution of label on the unfolded cortical surface resulting from an injection of 0.6 μCi of [³H]-proline into A I. Star indicates injection site. Parallel vertical lines delimit the heavily labeled region surrounding the injection site. Shaded areas bordered by continuous lines delimit areas containing transported label. Shaded region surrounded by an interrupted line near the border between A I and A II represents a projection deep in the sulcus. Oblique parallel lines on the rostral bank of the posterior ectosylvian sulcus represent a rostral branch of the sulcus. See Fig. 1.2 for further details (from Imig and Reale, 15).

injected into A I near an electrode penetration, in which neurons with best frequencies near 7.5 kHz were encountered. The star shows the location of the injection site and parallel lines delimit the heavily labeled area around the injection site in which silver grains are clustered over cell bodies. Transported label was found in stippled regions of fields A, P and VP and in a small region near the ventral border of field A I, deep within a rostral branch of the posterior ectosylvian sulcus. Most neurons located in the labeled area in field P had best frequencies between 7 and 8 kHz. Only one electrode penetration entered the labeled region in field A and here neurons had best frequencies of 5.0 and 7.7 kHz. Clearly, best frequencies of neurons near the injection site correspond closely with best frequencies of neurons in regions of fields A and P containing transported label.

Figure 1.6 illustrates an experiment in which isotope was injected into the low best-frequency representation of A I on the caudal bank of the posterior ectosylvian sulcus. The heavily labeled zone around the injection site was elongated along the penetration path of the injection pipet and covered the frequency representation between 0.8 and 1.5 kHz. Two patches of transported label were found on the caudal bank of the posterior ectosylvian sulcus. The dorsal elongated patch courses along the 1 kHz isofrequency strip in field P. The ventral patch probably corresponds in location with the low best-frequency representation of field VP. A small area was labeled on the anterior ectosylvian gyrus in the low best-frequency representation of field A. A small area of labeling is also seen near the fundus of the anterior ectosylvian sulcus. On the rostral bank of the posterior ectosylvian sulcus two small patches of label were found in the low-frequency region of field A I. Thus, the location of the injection site in the low-frequency representation in A I corresponds with the location of transported label in low-frequency representations in target fields A, P and A I.

Finally, isotope was injected at three points in the 20–25 kHz best-frequency representation in Fig. 1.7, which in this brain was located unusually near the posterior ectosylvian sulcus. Transported label was found in the high best-frequency representation common to fields P and VP and in an area dorsal to the injection site, which may belong to field A. In each case, best frequencies of neurons located near the injection site in A I correspond closely with best frequencies of neurons located in regions of fields A and P containing transported label.

Isotope was injected into other fields as well in other experiments. Figure 1.8, for example, shows the results of an experiment, in which isotope was injected into the high-frequency representation common to fields P and VP. Transported label was found in a dorsoventrally oriented band located along the high-frequency bor-

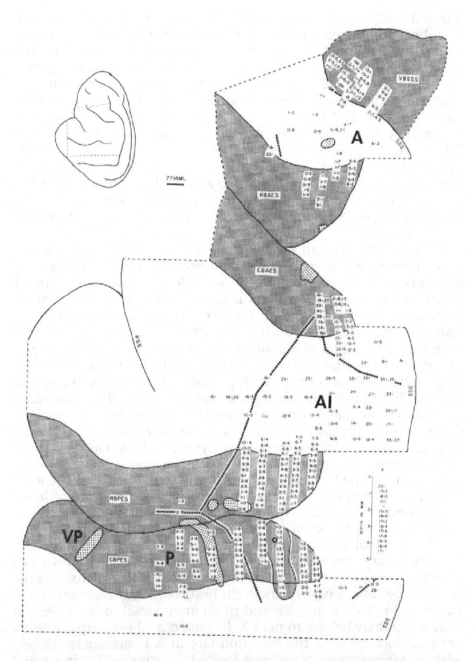

FIG. 1.6. Topographic distribution of label on the unfolded cortical surface resulting from an injection of 1 μCi of [³H]-proline into A I. The best-frequency sequence labeled with a small capital A (lower right quadrant) was encountered during an electrode penetration into the lower bank of the suprasylvian sulcus. The electrode entered the cortex at the location marked by the small capital letter A. See Figs. 1.2 and 1.5 for further details (from Imig and Reale, 15).

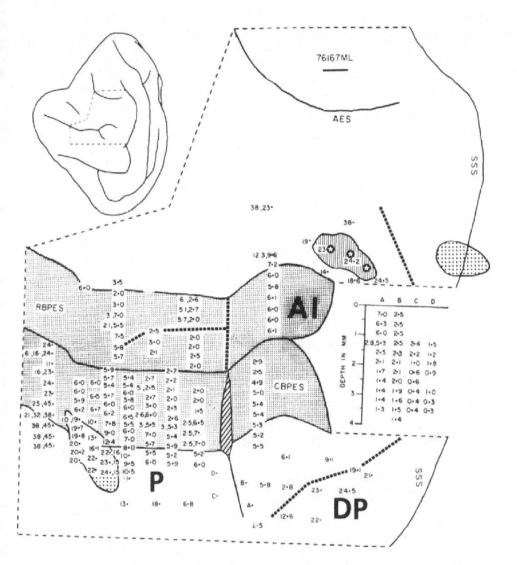

FIG. 1.7. Topographic distribution of label on the unfolded cortical surface resulting from three injections containing a total of 2 μCi of [^{3}H]-proline into A I. Locations of electrode penetrations into the banks of a caudal branch of the posterior ectosylvian sulcus are labeled A, B, C and D. Sequences of best frequencies encountered during these penetrations are shown in the inset in the lower right quadrant of the figure. See Figs. 1.2 and 1.5 for further details (from Imig and Reale, 15).

der between fields A I and A. Additional patches of label were also seen on the posterior ectosylvian gyrus, the rostral bank of the posterior ectosylvian sulcus and in a region ventral to A I. Again, best frequencies of neurons near the injection site are found to correspond closely to best frequencies of neurons in labeled regions of

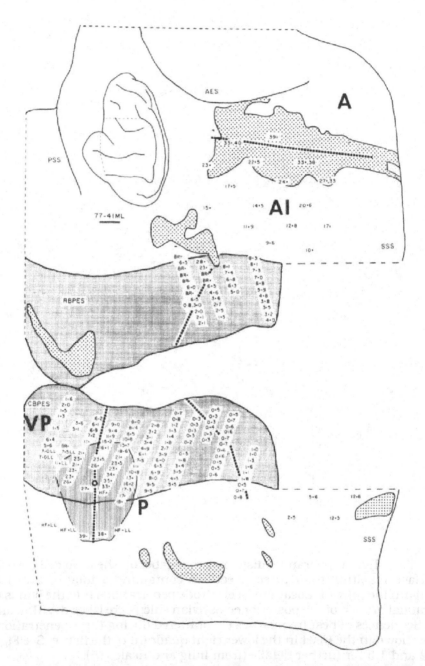

FIG. 1.8. Topographic distribution of label on the unfolded cortical surface resulting from an injection of 3.75 μCi of [³H]-proline near the border between the posterior and ventroposterior fields. See Figs. 1.2 and 1.5 for further details (from Imig and Reale, 15).

target fields. The results of the experiments illustrated in Figs. 1.7 and 1.8 also furnish anatomical confirmation of the specificity of interconnections between high-frequency representations on the middle and posterior ectosylvian gyri that Downman et al. (4) observed in physiological experiments.

Data obtained from experiments in which the isotope was injected into A I are summarized graphically in Fig. 1.9A. For each projection, a rectangle shows the range of best frequencies represented in the heavily labeled region surrounding the injection site (horizontal sides) and the range of best frequencies represented in regions containing transported isotope in a target field (vertical sides). Each rectangle is labeled to indicate the cortical field in which the transported label was found. For example, the rectangle labeled P in the lower left corner of the graph was obtained from the experiment illustrated in Fig. 1.6. Best frequencies represented in the heavily labeled region surrounding the injection site ranged between 0.5 and 1.8 kHz. Best frequencies represented in the portion of field P containing transported isotope ranged between 0.5 and 1.8 kHz. Subscript c designates target fields located in the hemisphere contralateral to the injection site. Similar graphs were prepared to summarize the results of injections that were made into fields P and VP (Fig. 1.9B) and into field A (Fig. 1.9C). In these graphs, each rectangle lies on the diagonal of the graph indicating that a common range of best frequencies is represented at the sources and terminations of corticocortical connections in fields A I, A, P and VP. Finally, the results of all injections into fields A I, A, P and VP are summarized in Fig. 1.9D. The abscissa represents the geometric mean of the best frequencies located within the heavily labeled region surrounding the injection site and the ordinate represents the geometric mean of best frequencies represented in labeled areas of a target field. Points on this graph cluster about the diagonal of the graph indicating that corticocortical projections interconnect similar portions of the best frequency representations in fields A I, A, P and VP in both hemispheres.

2.2. Topography of Projections: Patches and Bands

Ipsilaterally, each of fields A, A I, P and VP appears to be reciprocally connected with the other three (15). In general, projections between adjacent fields are more heavily labeled than projections between nonadjacent fields. Projections between fields A and VP, which are separated by the greatest distance, are only faintly labeled. Neurons in each of the four fields also project to multiple areas within the peripheral auditory belt. Within the contralateral

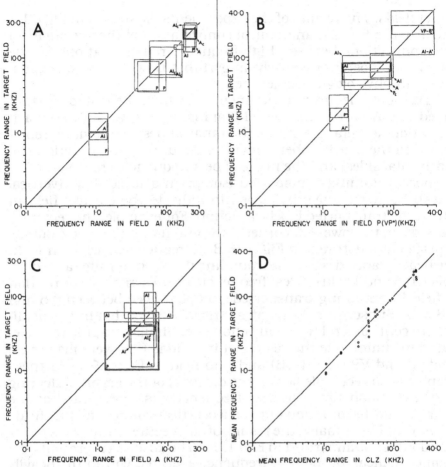

FIG. 1.9. Graphs showing the relationship between the range of best frequencies represented in the heavily labeled region around the injection site and the range of best frequencies represented in labeled regions of target fields A, A I, P or VP. Injection sites were located in field A I (A), fields P and VP (B) or in field A (C). In A, B, and C, each rectangle represents a projection from the injection site to a target field. The horizontal sides show the range of best frequencies in the heavily labeled region around the injection site (abscissa); the vertical sides show the range of best frequencies in the labeled portion of a target field (ordinate). Letters associated with each rectangle identify target fields. Subscript c indicates projection to fields in the hemisphere opposite the injection site. Asterisks indicate injection sites, which straddle fields P and VP (high-best frequencies) or are confined entirely to field VP (low-best frequencies). D, graph showing the relationship between the geometric mean of best frequencies represented in the heavily labeled zone around the injection site (CLZ) and the geometric mean of best frequencies represented in labeled regions of target fields for projections between fields A, A I, P and VP. Each point on the graph represents the center of gravity of a rectangle in A, B or C (from Imig and Reale, 15).

hemisphere, each of fields A, A I, P and VP is connected most strongly with the homotopic cortical region and additionally, some fields maintain connections with heterotopic cortical regions.

Although a single isotope injection into either A I, A, P, VP or A II results in dense labeling of a single cortical region centered at the injection site, multiple patches of transported label representing concentrations of axon terminals are generally found within each of several target areas. An example of this discontinuous pattern of labeling is seen in the projection from A I to field A. Figure 1.10 illustrates an autoradiograph of a tissue section cut in an oblique horizontal plane that passes through the dorsal tips of the anterior and posterior ectosylvian sulci. Four labeled bands formed by dense aggregates of silver grains cross perpendicular to the cortical laminae. Some of these bands extend from the cortical surface to the white matter. Areas containing lower concentrations of silver grains separate the bands from each other in the supra-granular layers, but bands may fuse in deeper layers. Patchy patterns of corticocortical terminations have been described in the somatic motor (1, 5, 16, 18), visual (21, 22, 26), auditory (3, 5, 7, 8, 14, 15) and other cortical areas (9) in monkeys and carnivores suggesting that this patchy pattern is a rather general property of corticocortical connectivity.

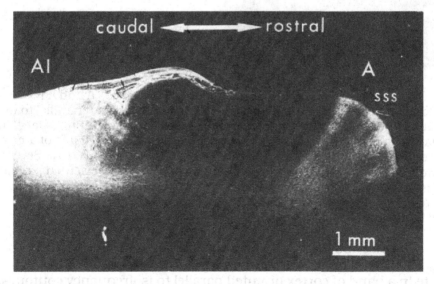

FIG. 1.10. Dark-field photograph showing the distribution of label in the anterior auditory field resulting from an injection of 50 μCi of [³H]-proline in the 1 kHz representation of ipsilateral A I. The tissue section was cut in an oblique horizontal plane perpendicular to the surface of the ectosylvian gyri.

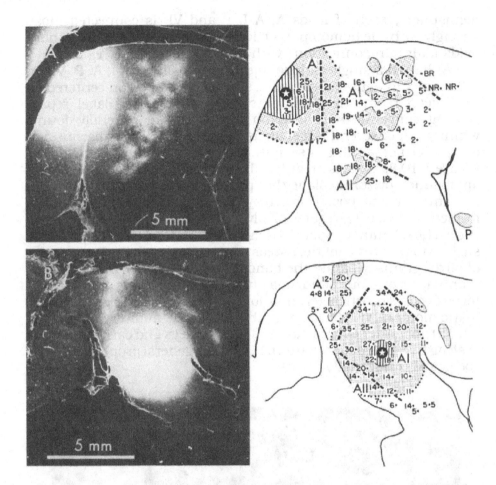

FIG. 1.11. Topography of projections between fields A and A I. Dark-
field photographs of autoradiographs of tissue sections cut parallel to the
flattened cortical surface appear on the left and the distribution of label in
relation to best frequency maps appears on the right. A, 16 μCi of a mixture of [³H]-leucine and [³H]-proline was injected into field A. Several
patches of label distributed throughout a band of cortex oriented parallel
to isofrequency contours are seen in A I. B, 3 μCi of [³H]-proline was injected in A I and a wedge of label is visible in field A. See Figs. 1.2 and 1.5
for further details.

   The distribution of label within a target field suggests that corticocortical fibers arising near an injection site spread out to terminate in a band of cortex oriented parallel to isofrequency contours.
One example is illustrated in Fig. 1.11A. On the left is a dark field
photograph of a tissue section cut parallel to the flattened cortical
surface and on the right is a drawing of the tissue section that has
imposed upon it the best frequency map obtained in this experi-

ment. The isotope was injected into the 6 kHz best-frequency representation of field A on the anterior ectosylvian gyrus. A dense region of labeling is apparent around the injection site. Within A I patches of transported label are found distributed along a band of cortex oriented parallel to isofrequency contours. Figure 1.11B illustrates a projection from A I to field A. In this case the isotope was injected into the 20 kHz representation of A I and a wedge of transported label was found within the 20 kHz representation of field A. The long axis of this wedge corresponds in orientation with the orientation of isofrequency contours in field A (17, Fig. 4). In each case, it is clear that corticocortical fibers diverge to terminate in an elongated band of cortex oriented parallel to isofrequency contours. Similar patterns of divergence were seen in many of the projections between fields A, A I, P and VP in both hemispheres (15).

3. Binaural Organization of A I

In addition to a tonotopic organization, there is a binaural organization within the cortical auditory system. Neurons within the brain stem auditory nuclei receive bilateral input and their projection upon higher centers ultimately carries binaural information to the cortex. Within auditory cortex, most neurons appear to be sensitive to binaural stimulation (2, 10). In a recent study (13), single neurons and neuron clusters were classified with respect to their monaural and binaural sensitivities at best frequency. This classification was based on the experimenters' judgments of the relative size of the neuronal response as viewed on the oscilloscope and heard over the audiomonitor, i.e., the number of spikes from an isolated single neuron or the amplitude of the response envelope of a neuron cluster to various monaural and binaural stimuli. With the exception of a few neurons responsive only to binaural stimulation, most neurons are excited by monaural stimulation. Some neurons are excited by monaural stimulation of one ear (usually the contralateral ear), but cannot be aroused by stimulation of the other ear. Other neurons are excited by stimulation of each ear. For a single neuron or cluster of neurons, that ear, stimulation of which produces the largest response at 70 dB SPL or below, is referred to as the dominant ear. Single neurons and neuron clusters, for which neither ear appeared particularly dominant, are referred to as equidominant.

The most commonly encountered binaural interaction was one in which the response to binaural stimulation was greater in size than the response to stimulation of either ear alone at the same fre-

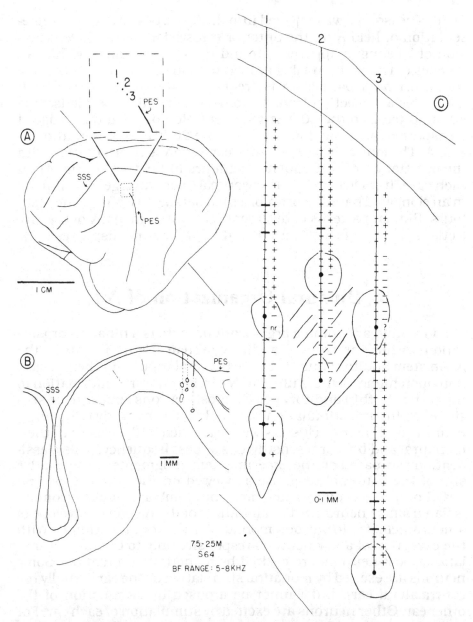

FIG. 1.12. Sequences of binaural interactions encountered during three electrode penetrations into A I. A, penetration locations near the dorsal tip of the posterior ectosylvian sulcus are shown on a drawing of the mapped hemisphere. The lines rostrodorsal and caudoventral to the hemisphere indicate the plane of tissue sections. B and C, outline drawing of a tissue section shows reconstructions of electrode tracks and marking lesions. Lines near marking lesions in C indicate the orientation of radial cell columns. Each depth at which a binaural interaction was assessed is

quency and sound pressure level. About two-thirds of the neurons studied displayed this characteristic, which was referred to as *summation*. Neurons that were excited less by binaural stimulation than by stimulation of the dominant ear alone at the same frequency and sound pressure level were classified as displaying *suppression*. In the vast majority of these cases, the contralateral ear was dominant and the neurons' responses were classified as contralateral dominant suppression. Stimulation of the ipsilateral ear alone was often ineffective in producing a response even at the highest sound pressure levels employed. Suppression was exhibited by about one-third of the neurons studied. About 5% of the neurons encountered displayed other types of responses.

When an electrode is advanced perpendicular to the cortical surface, most neurons located along the entire path display similar binaural interactions. On the other hand, nonperpendicular penetrations often pass from a zone in which neurons exhibit one binaural interaction to a zone in which neurons exhibit a different binaural interaction. This suggests that neurons with similar binaural properties are organized into columns. Figure 1.12 illustrates the results of an experiment in which a row of three electrode penetrations was placed near the dorsal tip of the posterior ectosylvian sulcus (Fig. 1.12A). Reconstructed electrode tracks are shown in drawings of a tissue section in Fig. 1.12B,C. The plane of section is indicated by lines above and below the hemisphere in Fig. 1.12A. In Fig. 1.12C, responses displaying similar binaural interactions are grouped at sequential points along each electrode penetration. In penetrations 1 and 3, the electrode encountered three distinct groupings of binaural responses. A series of summation (+) responses followed by a series of contralateral dominant suppression responses (−) and finally another series of summation responses were encountered. In penetration 2, a series of summation responses was followed by a series of contralateral dominant suppression responses. Lesions placed near the borders between different groups of summation and contralateral dominant suppression responses delimit a region in which only contralateral dominant suppression responses were encountered. Lesions lying on these borders are closely aligned with radial cell columns. Re-

---

FIG. 1.12. (continued)
marked by a horizontal bar. Heavy bars indicate locations at which single neurons were studied. Filled circles indicate the positions of marking lesions, whose outlines are also shown. With few exceptions, the electrode was advanced 50 μm between successive measurements (from Imig and Adrian, 13).

gions in which neurons display the same binaural interactions are referred to as binaural columns. In this experiment, penetrations 1 and 3 passed entirely through a contralateral dominant suppression column, which is flanked on both sides by summation columns.

Although experiments of the sort illustrated in Fig. 1.12 clearly demonstrated the existence of binaural columns within A I, they gave no information regarding the topography of these columns. To get at this question, closely spaced, nearly vertical penetrations were made into the cortex. Best frequencies, binaural interactions and aural dominances were assessed at two or three depths in each penetration. Figure 1.13 shows the results of one experiment in which a 1.5 mm × 4 mm area of A I was mapped. Best frequencies

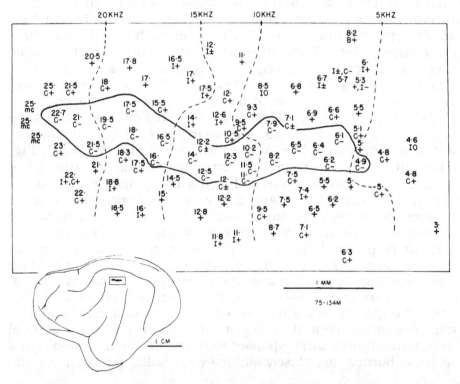

FIG. 1.13. Topographic distribution of binaural interactions, best frequencies and aural dominances of neuron responses. Isofrequency contours are indicated by interrupted lines. The hemisphere drawing shows the portion of the auditory cortex that was mapped. A continuous line delimits the borders of a contralateral dominant suppression column. If neuron response characteristics changed as a function of depth in a penetration, both responses are shown on the map separated by a comma or by the symbol ± (from Imig and Adrian, 13).

in the mapped region ranged between 3 kHz caudally and 25 kHz rostrally. In this experiment, an entire contralateral dominant suppression column was apparently mapped. This column occupies a rostrocaudally oriented strip of cortex. The long axis of this strip is oriented approximately orthogonal to isofrequency contours (shown by interrupted lines). The column is approximately 0.5 mm in width and nearly 4 mm in length and contains an orderly tonotopic representation of best frequencies between 4.9 and 22.7 kHz. Neurons encountered in the territory immediately surrounding the contralateral dominant suppression column all exhibited summation with the exception of those located at the rostral border. Three penetrations into this region encountered monaural contralateral (mc) neurons that were excited only by stimulation to the contralateral ear, but were not influenced by simultaneous stimulation to the ipsilateral ear. The finding that at least some binaural columns occupy strips of cortex that are oriented orthogonally with respect to isofrequency contours in A I has recently been confirmed by Middlebrooks et al. (20).

4. Corticocortical Connections Related to Binaural Maps

4.1. Patterns of Callosal Innervation in A I

Corticocortical projections to A I terminate within several densely labeled patches separated from each other by areas of lighter labeling. These patches often bear a striking and systematic relationship to binaural columns. The relationship of callosal connections to binaural columns in A I has been studied most extensively.

It has been possible to relate the pattern of callosal fiber connectivity to binaural columns by combining in the same experiment [3H]-proline injections with microelectrode mapping of A I (3, 14). In one experiment, 20 μCi of 3H-proline was injected into each of 35 locations in A I of the right hemisphere in order to label callosal fibers originating throughout the high-frequency representation. Injection sites were spaced at 1 mm intervals in a rectangular array covering the dorsal portion of the middle ectosylvian gyrus. Figure 1.14 shows an autoradiograph of a coronal section through A I in the hemisphere opposite the injection site. A series of dense patches of labeling and the track of an electrode, which passed nearly parallel to the cortical laminae, are clearly visible. Alternate series of contralateral dominant suppression responses (−) and summation responses (+) reflect the passage of the electrode

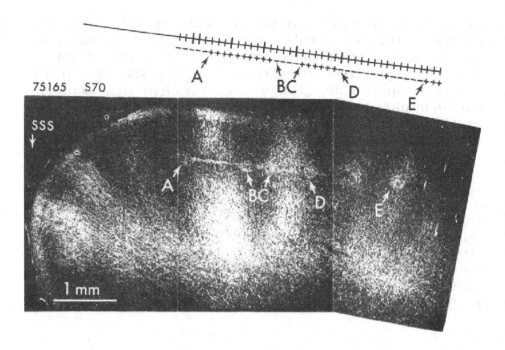

FIG. 1.14. Relationship between binaural and callosal columns seen in an autoradiograph of a coronal section through A I. Twenty µCi of [^{3}H]-proline was injected at each of 35 sites in A I of the right hemisphere. Six days later, response properties of neurons were studied at 100-µm intervals during an electrode penetration that was oriented nearly parallel to the cortical laminae of A I of the left hemisphere. Marking lesions (labeled A through E) placed at the borders of binaural columns are shown in both the autoradiograph and the sequence of binaural response symbols. Light vertical bars indicate responses of neuron clusters. Heavy bars indicate responses of single neurons. Dorsal is toward the left. Best frequencies in this penetration dropped from 18 kHz near the beginning to 12 kHz near the end (from Imig and Brugge, 14).

through alternate binaural columns. The positions of marking lesions are labeled by the letters A through E in the photomicrograph and in the sequence of binaural symbols. Inspection of the autoradiograph reveals that binaural columns lie in register with densely and sparsely labeled patches. Summation columns are more heavily labeled than contralateral dominant suppression columns, suggesting that summation columns receive more dense callosal innervation than do contralateral dominant suppression columns. The relationship of the pattern of labeling to binaural columns was not simply a fortuitous consequence of nonuniform labeling of the sources of callosal fibers in the opposite hemisphere as a result of the multiple isotope injections, because, in other experiments, patches of callosal fiber terminals labeled by a single injection of isotope into the opposite hemisphere displayed an identical correspondence to binaural columns.

Although the topography of binaural columns has not been described in detail, at least some columns appear to be elongated strips that cross several octaves of the frequency representation (Fig. 1.13). If the callosal innervation pattern is related to binaural columns in the manner just described, then one might expect to find a similar topographic pattern formed by callosal axon terminals. The topographic pattern of callosal innervation is best appreciated in tissue sections cut parallel to the flattened cortical surface. Figure 1.15 shows dark field photographs of tissue sections from the right hemispheres of two brains that were cut in this plane. In the left section, callosal fiber terminals were labeled by injecting isotope at multiple sites in A I of the opposite hemisphere. In the right section, degenerating axon terminals were stained using the Fink-Heimer technique after section of the corpus callosum (6). In both cases, light areas represent high concentrations and dark areas represent low concentrations of callosal axon terminals. Although the topographic distribution of callosal axon terminals is complex, certain consistent features are seen. Two prominent elongated bands composed of high concentrations of callosal terminals (α and β) run in a caudoventral to rostrodorsal direction parallel to the low-to-high best frequency gradient in A I. Caudal to α and β, callosal axon terminals are more uniformly distributed. The dorsal border of α is somewhat irregular and one or more branches extend toward and may enter the suprasylvian sulcus. Within α, callosal terminals are more densely concentrated than anywhere else in A I. In the region labeled gamma (γ), callosal terminals are more uniformly distributed than above, but the patchy distribution of these terminals is still evident. In spite of many individual variations in the topographic distributions of callosal terminals in different brains, the similarities between them remain impressive.

The pattern of callosal innervation in Fig. 1.15A corresponds closely with the map of binaural responses obtained in the same hemisphere (Fig. 1.16). Lesions A, C–F, H and K were each placed on borders between binaural columns and each corresponds closely with the border between a densely and a sparsely labeled region. Within densely labeled regions, summation ($+$) responses were generally encountered. Three electrode penetrations entered area α. All neurons encountered in this region displayed summation with the exception of those located at lesion G, which displayed ipsilateral dominance and suppression ($I-$). In other experiments, neurons displaying ipsilateral dominance and suppression are also found in densely labeled regions. Within β and γ all neurons dis-

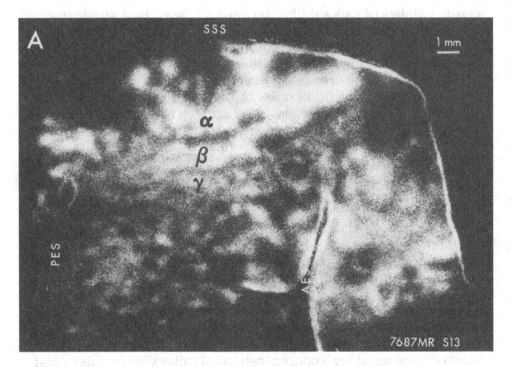

FIG. 1.15. Topographic distribution of callosal axon terminals seen in tissue sections cut parallel to the cortical surface, which had been flattened by gently pressing it onto a glass plate. A, autoradiograph of a tissue section in which callosal axon terminals were labeled with [³H]-proline. One day prior to sacrifice, 20 μCi of [³H]-proline was injected into each of

played summation. Within sparsely labeled regions neurons displayed contralateral dominance and suppression except near lesion H, where monaural contralateral responses were obtained. In other experiments, monaural contralateral responses were also encountered in sparsely labeled regions. Thus, these results provide further evidence that regions in which summation or ipsilateral dominant suppression responses are encountered receive more dense callosal innervation than regions in which contralateral dominant suppression or monaural contralateral responses are encountered.

In order to study the distribution of A I neurons that give rise to callosal axons, the HRP technique of retrograde axoplasmic transport was utilized. In three experiments, a solution of HRP was injected at multiple locations into A I. Following histological processing, cells that contained granules of HRP reaction product were found in A I contralateral to the injected hemisphere. Of the 3659 labeled cells counted in layers III through VI, 94% were pyramidal

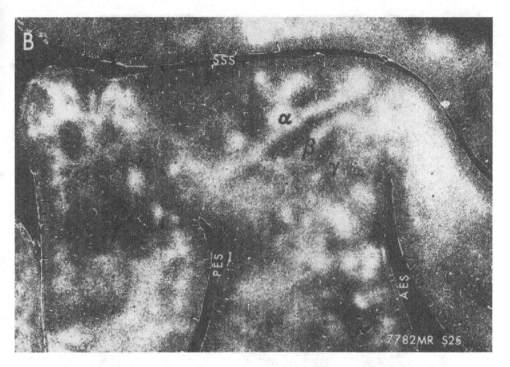

the 37 sites of the left hemisphere. B, photomicrograph of degenerating callosal axon terminals that were stained using the procedure described by Fink and Heimer (6). One week prior to sacrifice, the corpus callosum was completely sectioned using sterile surgical techniques (from Imig and Brugge, 14).

cells located in layers III and IV. The remaining 6% of the labeled cells were found in layers V and VI.

One striking feature of the labeled neurons in layers III and IV is their nonuniform distribution. Often labeled cells aggregate into clusters, which suggests that labeled cells might be more densely distributed in certain binaural columns than others. To test this hypothesis, injections of HRP into A I of one hemisphere were combined with electrophysiological mapping of binaural columns in the other. Following histological processing, the numbers of neurons in a narrow column of cortical tissue, the sides of which were parallel to radial cell columns, were counted. We refer to these tissue columns as sectors. One sector is outlined by interrupted lines in the tissue section drawing in the left panel of Fig. 1.17.

The histograms in Fig. 1.17 illustrate the relation between cell counts per sector and the binaural response characteristics of neurons located within those sectors. Each bin in the histogram repre-

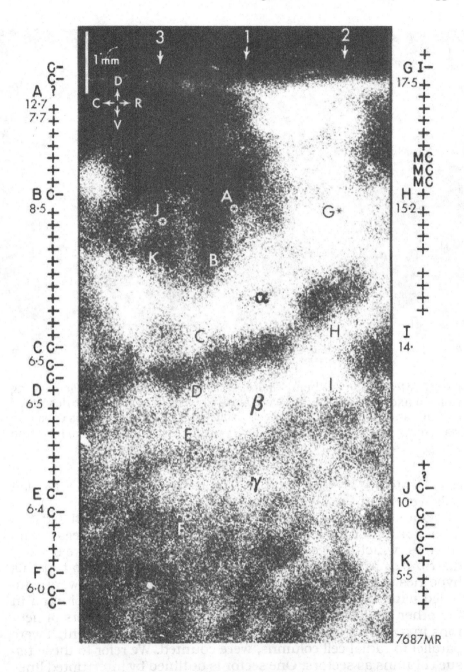

FIG. 1.16. Relationship between binaural and callosal columns in an autoradiograph of a tissue section cut parallel to the flattened cortical surface of A I. Following [³H]-proline injections in the left hemisphere (see legend of Fig. 1.15 for details), the responses of neurons were studied at 100-μm intervals in penetrations oriented parallel to the cortical laminae

sents the number of labeled cells in one sector and adjacent bins represent adjacent sectors. Below each histogram is plotted the sequence of binaural responses obtained in an electrode penetration, which was oriented nearly parallel to the cortical laminae. The electrode passed through each sector, from which the cell counts in the histogram were obtained. Binaural properties are plotted in register with absissae of the histograms using the marking lesions as reference points for both sets of data. The outlines of marking lesions and the locations of labeled cells are represented in the drawings below each histogram.

The nonuniform distribution of labeled cells in layers III and IV is reflected as peaks and valleys in the histograms. For ease of relating cell counts to binaural response properties, bins representing sectors located in contralateral dominant suppression columns have been shaded. Sectors that cross binaural column borders (e.g., the fifth bin from the right in the left panel of Fig. 1.17) are partially shaded. In Fig. 1.17 (left panel), the electrode passed through three summation columns (series of plusses) and each column corresponds with a peak in the histogram. Likewise, each of the four contralateral dominant suppression columns (series of minuses) penetrated by the electrode corresponds with a valley in the histogram. The results from a second penetration in the same experiment are illustrated in the right panel of Fig. 1.17. Between lesions C and E, the tissue section is cut obliquely with respect to radial cell columns. Consequently, in this region, cell counts were not obtained from sectors, but instead from tissue columns, whose sides were not parallel to radial cell columns. Because in this instance cell counts do not represent the number of labeled cells in the radial column from which binaural responses of neurons were obtained, we restrict consideration to the portion of the tissue section to the left of lesion C, which was cut parallel to the radial alignment of cells. Here, the peak in the histogram corresponds with a summation column and the valleys that flank the peak correspond with contralateral dominant suppression columns. The results obtained from both penetrations suggest that sectors located in sum-

---

FIG. 1.16. (continued)

in A I of the right hemisphere. The locations of marking lesions (labeled A through K), placed during three electrode penetrations (labeled 1, 2 and 3), are shown in both the autoradiograph and the sequences of binaural response symbols. Best frequencies (in kHz) obtained at points, where lesions were placed, are indicated beneath letters in the symbol sequences (from Imig and Brugge, 14).

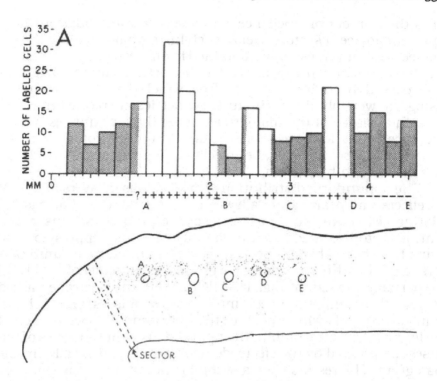

FIG. 1.17. Numbers and distributions of A I neurons giving rise to callosal fibers related to binaural properties of neurons. HRP (0.5 μL of a 40% solution) was injected into each of 36 locations in A I of the right hemisphere. Later, binaural responses of neurons were studied at 100-μm intervals during electrode penetrations that were oriented nearly parallel to the cortical laminae in A I of the left hemisphere. Marking lesions (labeled A through E) were placed near binaural column borders. Tissue sections were cut in the coronal plane. Dorsal is toward the left. Each labeled cell is represented as a dot in each drawing. Sectors have cross sectional dimensions of 180 × 200 μm. Borders of one sector are indicated by interrupted lines in panel 1.17A. Each histogram bin represents the number of labeled cells in a sector. Shaded bins represent sectors containing contra-

mation columns contain greater numbers of callosal projection neurons than do sectors located in contralateral dominant suppression columns.

Histograms in Fig. 1.18 illustrate the results obtained from another experiment and relate counts of labeled cells to binaural columns. As seen in Fig. 1.17, peaks and valleys in the histograms reflect the nonuniform distribution of labeled cells in layers III and IV. Cell counts obtained from sectors containing contralateral dominant suppression (−) or monaural contralateral (MC) neurons are represented by shaded bins in the histograms. Contralateral domi-

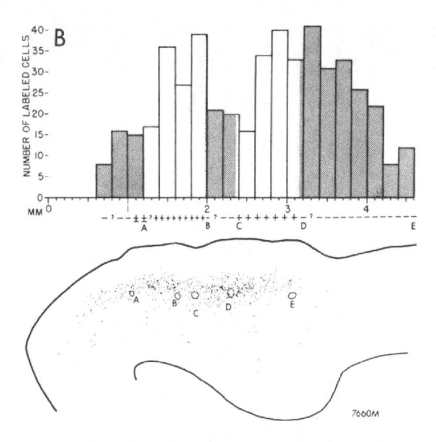

FIG. 1.17. (continued)
lateral dominant suppression neurons. Open bins represent sectors containing summation neurons. In the penetration illustrated in A, best frequencies ranged from 16 kHz near the beginning of the penetration to 10 kHz at the end. The penetration illustrated in B was located about 1 mm posterior to that illustrated in A and best frequencies ranged from 16.5 kHz near the beginning of the penetration to 7.4 kHz at the end (from Imig and Brugge, 14).

nant suppression responses in general tend to correspond with valleys of the histograms and examples are located near lesions A, B and D in Fig. 1.18A, near lesions B, C and D in Fig. 1.18B and near lesion C in Fig. 1.18C. A series of monaural contralateral (MC) responses to the right of lesion C in Fig. 1.18A also appears to correspond with a valley in the histogram. Cell counts obtained from sectors located in summation columns (responses indicated by plusses) are represented by open bins and these, in general, correspond with peaks in the histograms. Examples of these are located between lesions A and B, between lesions B and C in Fig. 1.18A, be-

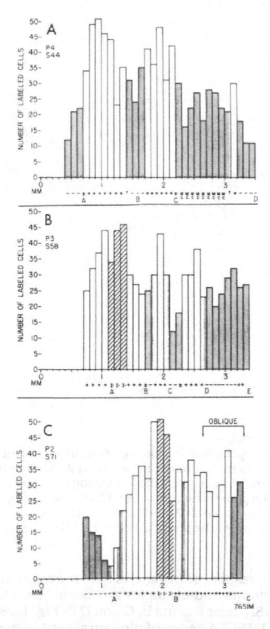

FIG. 1.18. Numbers and distributions of A I neurons giving rise to callosal fibers related to binaural properties of neurons. HRP (0.5 μL of a 33% solution) was injected into each of 35 locations in A I of the right hemisphere. Later, binaural properties of neurons were studied at 100 μm intervals during electrode penetrations oriented nearly parallel to the cortical laminae in A I of the left hemisphere. Positions of marking lesions (labeled A through E) are shown in the sequences of binaural response

tween lesions B and C and between lesions C and D in Fig. 1.18B and between lesions A and B in Fig. 1.18C. Furthermore, in two penetrations, ipsilateral dominant suppression (I−) responses were encountered. Histogram bins corresponding with these responses are distinguished by diagonal lines. Between lesions A and B in Fig. 1.18B and between lesions A and B in Fig. 1.18C, ipsilateral dominant suppression responses correspond with peaks in the histograms. In general, the results obtained in the experiments illustrated in Figs. 1.17 and 1.18 suggest that sectors, in which summation or ipsilateral dominant suppression neurons are found, contain more labeled cells than do sectors, in which contralateral dominant suppression or monaural contralateral neurons are found. These data also suggest that the sources of callosal axons may be more heavily concentrated in summation columns located dorsally in A I than those located more ventrally.

Careful inspection of the histograms in Figs. 1.17 and 1.18 reveals certain examples that do not fit the general relationship outlined above. Perhaps the most striking example is the summation response located at lesion A in Fig. 1.18C, which corresponds with the lowest bin value in the histogram, a finding counter to the general trend. Furthermore, neither the summation column near lesion E in Fig. 1.18B nor the suppression column near lesion B in Fig. 1.18C corresponds with a clear peak or valley in the histogram. Finally, even in regions where there is apparently a good correspondence between histogram peaks and valleys and binaural columns (e.g., between lesions A and C in Fig. 1.18A), some sectors in contralateral dominant suppression columns contain more labeled cells than sectors in summation columns.

FIG. 1.18. (continued)

symbols. Penetrations are oriented as in Fig. 1.17. Each histogram bin represents the number of labeled cells per sector. Shaded bins represent sectors containing contralateral dominant suppression and monaural contralateral neurons. Open bins represent sectors containing summation neurons. Bins with diagonal lines represent sectors containing ipsilateral dominant suppression neurons. Sectors had cross sectional dimensions of 180 × 100 μm at the III–IV border. Electrode penetrations were approximately parallel and spaced about 1 mm apart. The penetration illustrated in A was located most rostral, that in C most caudal. In A, best frequencies decreased from 18 kHz near the beginning of the penetration to 14.5 kHz near the end, in B from 16.5 to 9 kHz, and in C from 13 to 5 kHz. In C, OBLIQUE indicates that the tissue section in this region was cut obliquely with respect to radial cell columns (from Imig and Brugge, 14).

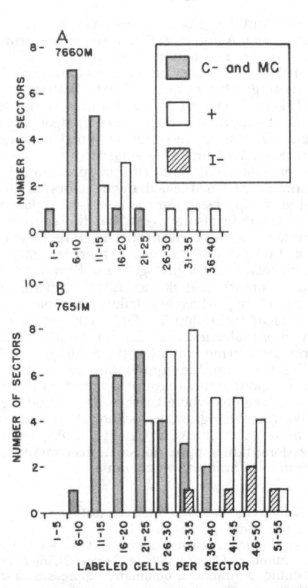

FIG. 1.19. Distributions of cell counts per sector related to binaural properties of neurons. Each histogram includes all sector values obtained in one experiment, except those from sectors in which the binaural properties of neurons varied or those from sectors that were located in regions of cortex cut obliquely with respect to radial cell columns. Shaded bins represent the distributions of cell counts obtained from sectors containing contralateral dominant suppression neurons (A) or contralateral dominant suppression and monaural contralateral neurons (B). Open bins represent the distribution of cell counts obtained from sectors containing summation neurons; diagonal lines designate bins representing the distribution of cell counts obtained from sectors containing ipsilateral dominant suppression neurons (from Imig and Brugge, 14).

Distribution of cell counts obtained from regions that contain neurons with similar binaural properties are graphically represented in Fig. 1.19. These distributions include all sector values obtained during penetrations in one experiment except those from sectors that crossed binaural column borders or were located in regions where the cortex was cut obliquely with respect to radial cell columns. In Figs. 1.19A and B, values from experiments 7660M and 7651M, respectively, are plotted. Shaded bins represent the distribution of cell counts obtained from sectors containing contralateral dominant suppression or monaural contralateral neurons. These values have been combined, since a *t*-test did not reveal any significant difference between the two populations. Open bins represent the distribution of cell counts obtained from sectors containing summation neurons. Although the two distributions overlap, values obtained from sectors containing summation neurons in general exceed values obtained from sectors containing contralateral dominant suppression and monaural contralateral neurons. A *t*-test shows that the distributions are significantly different ($p < 0.001$). Bars that contain diagonal lines designate the distribution of cell counts obtained in regions containing ipsilateral dominant suppression neurons. Although few in number, these sectors contain significantly more labeled neurons than do sectors in which summation neurons are found (Mann Whitney U-test; $p < 0.033$). In summary, sectors containing contralateral dominant suppression or monaural contralateral neurons contain significantly fewer labeled cells than do sectors containing summation neurons. The greatest numbers of cells are found in sectors containing ipsilateral dominant suppression neurons.

From the mean number of labeled cells per sector and the cross-sectional area of sectors, it is possible to estimate the number of labeled cells in a binaural column with a cross sectional area of one square millimeter. These computations reveal that fewer cells were labeled in experiment 7660M than in experiment 7651M. The number of labeled cells per square millimeter of summation column was estimated to be 614 in experiment 7660M and 1987 in experiment 7615M. The number of labeled cells per square millimeter of contralateral dominant suppression column was estimated to be 315 in experiment 7660M and 1161 in experiment 7651M. The differences in cell counts in these experiments may reflect differences in staining since tissue sections from experiment 7660M were less intensely stained for the HRP reaction product than were those from experiment 7651M. Although the technique employed here is not adequate to produce a reliable estimate of the absolute number of callosal projection neurons in different binaural columns, it may provide an estimate of the relative number of callosal projection neurons in different binaural columns. In experiment 7660M, sectors located in summation columns contained 1.95 times as many labeled neurons as did sectors that were located in contralateral dominant suppression columns. In experiment 7651M, 1.76 times as many labeled neurons were found in sectors containing summation neurons as in sectors containing contralateral dominant suppression and monaural contralateral neurons. Sectors in which ipsilateral dominant suppression neurons were found contained 2.18 times as many labeled neurons as did sectors in which contralateral dominant suppression and monaural contralateral neurons were found. Although the difference in staining of tissue sections from these two experiments may be reflected dramatically in the absolute number of cells that are labeled, the relative number of labeled cells in different binaural columns is more constant between experiments.

4.2. Patterns of Ipsilateral Corticocortical Terminations in A I

We have already shown that some ipsilateral corticocortical axon terminals are more densely aggregated in some regions of A I than in others (Figs. 1.10 and 1.11). In some cases, these dense patches of corticocortical terminals appear elongated in a direction parallel to the low-to-high gradient of the best frequency map, just as did the terminals of callosal fibers. Figure 1.20 illustrates patterns of labeling in A I that were obtained in four experiments. In A and B, isotope was injected into field A on the anterior ectosylvian gyrus, whereas, in C and D, isotope was injected into field P on the caudal bank of the posterior ectosylvian sulcus. In each case, tissue sections were cut parallel to the flattened gyral surfaces of the ectosylvian region. In A, several dense aggregates of silver grains appear to coalesce to form a prominent rostrocaudally elongated band that is flanked above and below by sparsely labeled bands. In B, three densely labeled bands are recognizable. Arrows mark a narrow band of label that runs from the posterior ectosylvian sulcus onto the surface of the middle ectosylvian gyrus, and two considerably broader bands that are located more dorsally. Above and below the elongated patches of label in A I, more complex patterns of labeling are seen. Neurons in field P may also terminate within regions elongated in the direction of the low-to-high gradient of the best frequency map in A I. In C, isotope was injected near a point in field P, in which neurons with best frequencies of 10 kHz were found. Two densely labeled patches (marked by arrows) in field A I are elongated along the low-to-high gradient of the best frequency map. A dense region of labeling, located ventral to these elongated patches, probably extends across the ventral border of A I. Labeling may also ex-

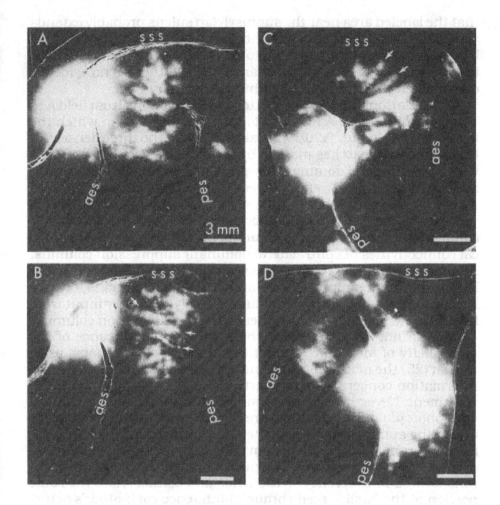

FIG. 1.20. Topography of projections from fields A and P upon A I as seen in dark-field photographs of tissue sections cut parallel to the flat-tened cortical surface. A mixture of [³H]-proline and [³H]-leucine was in-jected into each hemisphere illustrated in A, B, C and D. Quantities in-jected were 55 μCi in A, 44 μCi in B and 50 μCi in C and D.

tend across the dorsal border of A I. Thus, fibers arising in both fields A and P and projecting ipsilaterally to A I may terminate in bands that are elongated along the low-to-high best frequency gra-dient.

In the hemisphere opposite that illustrated in C, isotope was also injected into the 10 kHz representation of field P, but in this case, there was no indication of elongated patches of label in A I. A tissue section from this experiment is seen in D. On the middle ect-osylvian gyrus two densely labeled regions are seen. From the best frequency map, which was obtained in this experiment, it appears that the labeled area near the suprasylvian sulcus probably extends across the dorsal border of A I. The densest portion of the ventrally labeled region lies largely ventral to A I. Although lighter labeling extends from this region some distance into A I, there is no evidence of elongated patches of label within the primary field.

The relationship between patterns of projections from field A to binaural columns in A I was explored in the brain from which the tissue section in Fig. 1.20B was taken. Each of the three densely labeled elongated patches marked by arrows corresponded precisely with a contralateral dominant suppression column; lighter areas of labeling above, below and between these patches corresponded with summation columns. This is just opposite to the pattern of callosal labeling. Thus, it appears that the corticocortical connectivities of summation columns differ from the corticocortical connectivities of contralateral dominant suppression columns. Summation columns receive a denser callosal input than do contralateral dominant suppression columns. On the other hand, contralateral dominant suppression columns receive a denser input from field A in the ipsilateral hemisphere than do summation columns.

We do not yet understand the functional significance of the multiplicity of tonotopic representation in the cat. Perhaps, as in the bat (25) the neural components in each area function to extract information concerning some particular aspect of the acoustic environment. Nevertheless, it is clear that distributed among the several tonotopic representations are neurons that are linked via the ascending auditory pathway to the same limited portion of the basilar membrane. In turn, these same elements are linked via corticocortical connections into a coordinated network that allows neurons located in different cortical fields and connected to the same portion of the basilar memebrane to influence each other's activities.

From the information currently available, it is not possible to determine whether corticocortical projection fibers connect together areas of common binaural function in different auditory fields in a manner similar to their role connecting together cortical regions representing common portions of the basilar membrane. On the other hand, the relation between the pattern of callosal connectivity and the binaural representation in A I suggests that the callosal pathway may function to transmit a specific subset of information between the primary auditory fields. In experiments in which HRP was injected into A I, nearly twice as many labeled neurons were found in sectors containing summation or ipsilateral dominant suppression neurons as in sectors containing contralateral dominant suppression or monaural contralateral neurons. Thus, binaural information transmitted by the population of A I callosal cells differs from the binaural information transmitted by the general population of A I neurons. It is conceivable that each corticocortical pathway carries a different specific subset of information about the acoustic environment.

Abbreviations

A	Anterior auditory field
AES or aes	Anterior ectosylvian sulcus
A I	Primary or first auditory field
A II	Second auditory area
BR	Response over a broad frequency range at lowest-threshold intensity
C	Contralateral dominance
C− or −	Contralateral dominance and suppression
C +	Contralateral dominance and summation
CBAES	Caudal bank of the anterior ectosylvian sulcus
CBPES	Caudal bank of the posterior ectosylvian sulcus
CLZ	Central labeled zone—the heavily labeled region surrounding an isotope injection site
DP	Dorsoposterior area
HF	High (>25kHz) best frequency
I	Ipsilateral dominance
I−	Ipsilateral dominance and suppression
I+	Ipsilateral dominance and summation
Kcps	Kilocycles per sec (kiloHertz)
kHz	KiloHertz
LF	Low (<1 kHz) best frequency
LL	Long latency response (>50 ms)
M	Multiple best frequencies
MC or mc	Monaural contralateral
NR or nr	No response
P	Posterior auditory field
PES or pes	Posterior ectosylvian sulcus
PSS or pss	Pseudosylvian sulcus
RBAES	Rostral bank of the anterior ectosylvian sulcus
RBPES	Rostral bank of the posterior ectosylvian sulcus
SPL	Sound pressure level
SSS or sss	Suprasylvian sulcus
SW	Swish
T	Temporal auditory area
V	Ventral auditory area
VBSSS	Ventral bank of the suprasylvian sulcus
VP	Ventroposterior field

References

1. BOWKER,R. M., AND COULTER, J. D. Intracortical connectivity in somatic sensory and motor cortex of monkeys. In: *Cortical Sensory Organization*, Vol. 1, Chapter 8, edited by C. N. Woolsey. Clifton, New Jersey: Humana Press, 1981.

2. BRUGGE, J. F., DUBROVSKY, N. A., AITKIN, L. M., AND ANDERSON, D. J. Sensitivity of single neurons in auditory cortex of cat to binaural tonal stimulation; effects of varying interaural time and intensity. *J. Neurophysiol.*, 32: 1005–1024, 1969.

3. BRUGGE, J. F., AND IMIG, T. J. Some relationships of binaural response patterns of single neurons to cortical columns and interhemispheric connections of auditory area A I of cat cerebral cortex. In: *Evoked Electrical Activity in the Auditory Nervous System*, edited by R. F. NAUNTON AND C. FERNANDEZ. New York: Academic Press, 1978, pp. 487–503.

4. DOWNMAN, C. B. B., WOOLSEY, C. N., AND LENDE, R. A. Auditory areas I, II and Ep: Cochlear representation, afferent paths and interconnections. *Bull. Johns Hopkins Hosp.*,106: 127–142, 1960.

5. EBNER, F. F., AND MYERS, R. E. Distribution of corpus callosum and anterior commissure in cat and raccoon. *J. Comp. Neurol.*, 124: 353–366, 1965.

6. FINK, R. P., AND HEIMER, L. Two methods for selective silver impregnation of degenerating axons and their synaptic endings in the central nervous system. *Brain Res.*, 4: 369–374, 1967.

7. FITZPATRICK, K. A., AND IMIG, T.J. Auditory cortico-cortical connections in the owl monkey. *J. Comp. Neurol.*, 192: 589–610, 1980.

8. FITZPATRICK, K. A., AND IMIG, T. J. Organization of connections of primate auditory cortex. This volume, chapter 4.

9. GOLDMAN, P. S., AND NAUTA, W. J. H. Columnar distribution of cortico-cortical fibers in the frontal association, limbic and motor cortex of the developing rhesus monkey. *Brain Res.*,122: 393–413, 1977.

10. HALL, J. L., AND GOLDSTEIN, M. H., JR. Representation of binaural stimuli by single units in primary auditory cortex of unanesthetized cats. *J. Acoust. Soc. Amer.*,43: 456–461, 1968.

11. HIND, J. E. An electrophysiological determination of tonotopic organization in auditory cortex of the cat. *J. Neurophysiol.*,16: 473–489, 1953.

12. HIND, J. E., ROSE, J. E., DAVIES, P. W., WOOLSEY, C. N., BENJAMIN, R. M., WELKER, W., AND THOMPSON, R. F. Unit activity in the auditory cortex. In: *Neural Mechanisms of the Auditory and Vestibular Systems*, edited by G. L. RASMUSSEN AND W. F. WINDLE. Springfield, Illinois: Charles C Thomas, 1960, pp. 201–210.

13. IMIG, T. J., AND ADRIAN, H. O. Binaural columns in the primary field (A I) of cat auditory cortex. *Brain Res.*, 138: 241–257, 1977.

14. IMIG, T. J., AND BRUGGE, J. F. Sources and terminations of callosal axons related to binaural and frequency maps in primary auditory cortex of the cat. *J. Comp. Neurol.*, 182: 637–660, 1978.

15. IMIG, T. J., AND REALE, R. A. Patterns of cortico-cortical connections related to tonotopic maps in cat auditory cortex. *J. Comp. Neurol.*, 192: 293–332.
16. JONES, E. G., COULTER, J. D., AND HENDRY, S. H. C. Intracortical connectivity of architectonic fields in the somatic sensory, motor and parietal cortex of monkeys. *J. Comp. Neurol.*, 181: 291–348, 1978.
17. KNIGHT, P. L. Representation of the cochlea within the anterior auditory field (AAF) of the cat. *Brain Res.*, 130: 447–467, 1977.
18. KÜNZLE, H. Cortico-cortical efferents of primary motor and somatosensory regions of the cerebral cortex in *Macaca fascicularis*. *Neuroscience*, 3: 25–39, 1978.
19. MERZENICH, M. M., KNIGHT, P. L., AND ROTH, G. L. Representation of cochlea within primary auditory cortex in the cat. *J. Neurophysiol.*,38: 231–249, 1975.
20. MIDDLEBROOKS, J. C., DYKES, R. W., AND MERZENICH, M. M. Binaural response-specific bands in primary auditory cortex (A I) of the cat: topographic organization orthogonal to isofrequency contours. *Brain Res.*, 181: 31–48, 1980.
21. MONTERO, V. M. Patterns of connections from the striate cortex to cortical visual areas in superior temporal sulcus of macaque and middle temporal gyrus of owl monkey. *J. Comp. Neurol.*,189: 45–59, 1980.
22. MONTERO, V. M. Corticocortical connectivities of striate cortex in the cat. In: *Cortical Sensory Organization*, Vol. 2, Chapter 2, edited by C. N. Woolsey. Clifton, New Jersey: Humana Press, 1981.
23. REALE, R. A., AND IMIG, T. J. Tonotopic maps of auditory cortex in the cat. *J. Comp. Neurol.*,192: 265–292, 1980.
24. SINDBERG, R. M., AND THOMPSON, R. F. Auditory response fields in ventral temporal and insular cortex in cat. *J. Neurophysiol.*, 25: 21–28, 1962.
25. SUGA, N. Functional organization of the auditory cortex beyond tonotopic representation. This volume, chapter 6.
26. WONG-RILEY, M. Columnar corticocortical interconnections within the visual system of the squirrel and macaque monkeys. *Brain Res.*,162: 201–217, 1979.
27. WOOLSEY, C. N. Organization of cortical auditory system: a review and a synthesis. In: *Neural Mechanisms of the Auditory and Vestibular Systems*, edited by G. L. RASMUSSEN AND W. F. WINDLE. Springfield, Illinois: Charles C Thomas, 1960, pp. 165–180.
28. WOOLSEY, C. N. Organization of cortical auditory system. In: *Sensory Communication*, edited by W. A. ROSENBLITH. Cambridge: The MIT Press, 1961, pp. 235–257.
29. WOOLSEY, C. N., AND WALZL, E. M. Topical projection of nerve fibers from local regions of the cochlea to the cerebral cortex of the cat. *Bull. Johns Hopkins Hosp.*,71: 315–344, 1942.

Chapter 2

Auditory Forebrain Organization

Thalamocortical and Corticothalamic Connections in the Cat

Michael M. Merzenich,[1] Steve A. Colwell[1] and Richard A. Andersen[2]

Coleman Laboratory,[1] University of California, San Francisco, San Francisco, California, and the Department of Physiology,[2] The Johns Hopkins University, Baltimore, Maryland

1. Introduction

Studies conducted over the past five years have led (1) to a redefinition of the boundaries of auditory cortical fields in cats; (2) to a new understanding of the geometry and sources of input to given cortical fields; and (3) to a much more topographically refined definition of the destinations of projections originating in given auditory cortical fields. This new understanding of auditory thalamocortical and corticothalamic organization has been derived from

43

unit mapping studies and from results of combined (anterograde and retrograde) tracer injections in a physiological-anatomical study during which injections of tracer materials were introduced at functionally defined auditory field sites.

1.1. Redefinition of Auditory Cortical Fields in the Cat

Before discussing auditory thalamocortical and corticothalamic connections in the cat, it is necessary briefly to review our current understanding of the basic organization of auditory cortical fields and of medial geniculate body subdivisions. Microelectrode unit mapping studies have led to a picture of auditory field boundaries illustrated in Fig. 2.1. (1, 3, 14, 15) (also see Imig et al., chapter 1 this volume). Physiological mapping results were in general consistent with earlier results of Woolsey and colleagues (30–33). Boundaries of fields are drawn somewhat differently, with the appreciation in these unit mapping studies that there are reversals (not discontinuities) in representational sequences across all borders of adjoining topographic fields.

1.2. Internal Organization of Auditory Cortical Fields

We now recognize that there are at least four large "cochleotopic" (or "tonotopic") auditory cortical fields in cats, with a large region ventral to A I having no evident topography. This conclusion was predated by a similar conclusion in earlier studies using evoked response techniques (30–33). Within all identified topographically organized fields, there is a rerepresentation of the cochlear sensory epithelium across one field dimension (13, 14, 17, 18; also see Imig et al., chapter 1, this volume; see Figs. 2.1, 2.2). This pattern of organization was earlier described in detail for A I and a more anterior field, by Tunturi in the dog (27, 28). It was also manifested by the banded patterns of arrays of evoked response within A I, recorded by Woolsey and Walzl (33).

Imig and colleagues (5, 11, 12) provided initial evidence that this isorepresentational axis of A I is divisible into binaural subunits (which they called "binaural columns"), within which column-specific binaural neural response properties are recorded. Recent studies in this laboratory (19) along with anatomical–physiological studies of Brugge and Imig (5) have indicated that these "columns" in the higher frequency aspect of A I are actually bands, extending across A I roughly orthogonal to the axis of representation of frequency, and that there are bands in alternating

Cat Auditory Fields Internal Organization of A1

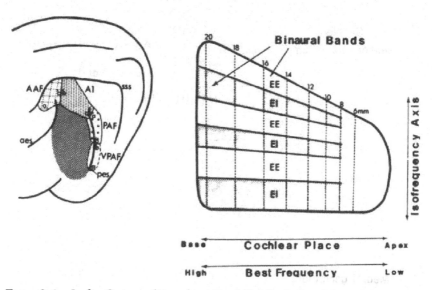

FIG. 2.1. Left: Cat auditory cortical fields (after Merzenich et al., 15). AAF = anterior auditory field; A I = primary auditory field; PAF = posterior auditory field; VPAF = ventral posterior auditory field. These four fields are strictly cochleotopically (tonotopically) organized (also see Imig et al., Chapter 1, this volume), with reversals in representation across the cochlear base (b) along the AAF–A I and PAF–VPAF boundaries, and across the cochlear apex (a) along the A I–PAF border. The crosshatched region is an auditory responsive zone in which no cochleotopic order has yet been defined. Aes, anterior ectosylvian sulcus; pes, posterior ectosylvian sulcus; sss, suprasylvian sulcus.

Right: Schematic illustration of the internal organization of A I. See text for description. Numbers represent cochlear isorepresentational lines (mm from apex). Redrawn from Merzenich et al. (15).

sequence, in which neurons have predominantly excitatory– excitatory (EE) response properties (driven responses to simultaneous stimulation of two ears are greater than those to either ear alone) and, in the alternate bands, excitatory–inhibitory (EI) response properties (stimulation of one ear drives the neuron, while the other ear inhibits the driven responses) (5, 19). There is growing evidence (in studies now underway in our laboratory) for band-specific differences in neural response properties in alert cats. It appears that these A I subunits of like sign (EI or EE "bands") have different (possibly different individual) functional significance and may represent end-processing regions for parallel subdivisions of the auditory projection system of the cat (16, 19).

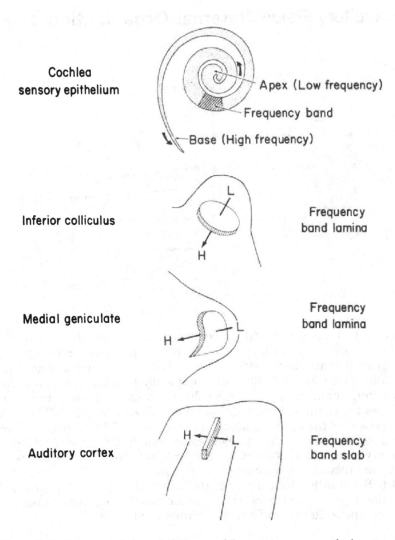

Cochlea
sensory epithelium

Apex (Low frequency)

Frequency band

Base (High frequency)

Inferior colliculus

L

H

Frequency
band lamina

Medial geniculate

H ← | ┐ L

Frequency
band lamina

Auditory cortex

H ←|→ L

Frequency
band slab

FIG. 2.2 Representation of the cochlear sensory epithelium within "main line" auditory nuclei and cortical fields. Any sector of the cochlea is represented across a relatively flat sheet of neurons that extends across the central nucleus of the inferior colliculus from edge to edge; across a folded sheet of neurons in the lateral part of the ventral division of the MGB; and across a roughly straight slab of neurons within A I (as well as in AAF, PAF, and VPAF). Representationally, there is a rerepresentation of any given cochlear locus across one dimension of A I, and across two dimensions of the central nucleus of the inferior colliculus and lateral ventral MGB. A representational dimension is lost via convergence from MGB neuronal sheets to A I isofrequency (isorepresentational) lines (see 14–16, 18).

1.3 Internal Organization of the Medial Geniculate Body (MGB)

Anatomical results described later are most consistent with definition of MGB subdivisions defined in Golgi studies of the nucleus (20, 21, 24), although they are not consistent in all respects with the details of descriptions of those studies. In subsequent descriptions, the terminology of Morest (20, 21) will be used in describing MGB organization.

2. Basic Approach

The goal of these studies was to define the arrays of thalamic neurons, in three dimensions, projecting to restricted A I loci. Also of interest was the relationship of the projecting neuronal arrays to the patterns of terminals of the descending corticothalamic projections.

Brief physiological mapping studies were conducted to define the approximate locations within A I at which combined injections of anterograde and retrograde tracer were introduced. In these physiological experiments, (a) the approximate boundaries of A I were defined, (b) the orientation of the isorepresentational frequency axes of A I were determined and (c) the "best" frequency positions were identified for single or multiple locus injection sites. Generally, in the multiple injection experiments, a second injection was made either at another location along the same "isofrequency contour" or at a different "best frequency" representational location in the same auditory field.

After 24–40 hour survival periods, animals were perfused with paraformaldehyde and the brains were processed using standard histochemical (DAB) and autoradiographic techniques. These procedures are described in detail in other reports from this laboratory (1–4, 7–8).

3. Summary of Results

3.1. Geometry of Arrays of Neurons Projecting from the MGB to A I

Arrays of neurons projecting from MGB subdivisions to restricted A I loci are illustrated by example in Figs. 2.3, 2.4. With a moderate sized injection (Fig. 2.3), the array constituted a continuous neuronal sheet that bisected the lateral part of the ventral nucleus,

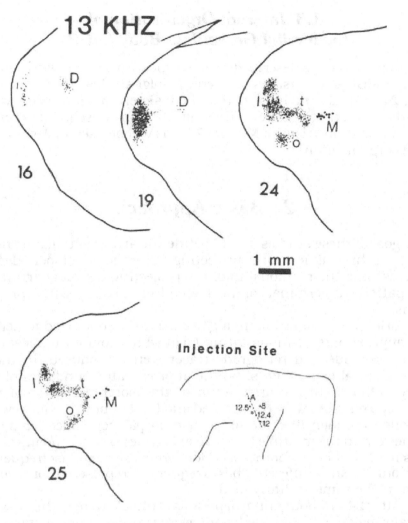

FIG. 2.3. Drawings of sections through the MGB in a cat, illustrating the geometric figure of the complex neuronal array that projects to a 13 kHz representational (horseradish peroxidase injection) site within A I. The HRP injection site is illustrated diagramatically at the lower right. These numbers represent best frequencies defined for neurons within penetrations normal to indicated sites. The approximate isorepresentational axis is indicated by the dashed line. This 0.2 μL injection (and all other illustrated injections) was wholly restricted within A I. Note the complex sheetlike folded array in the lateral (1), transitional (t) and ovoidal (o) parts of the ventral division, as well a projecting column of neurons within the deep part of the dorsal division (D) and a projecting cluster of neurons within the medial division (M). Numbered sections (S) were 150 μm apart. The deep dorsal array extended well rostral to S 16; the ovoidal–lateral arrays in the ventral division extended more than 450 μm caudal to S 25. Note the broken appearance of the array (banded, in three dimensions) in S.24 and 25.

7.5 KHZ

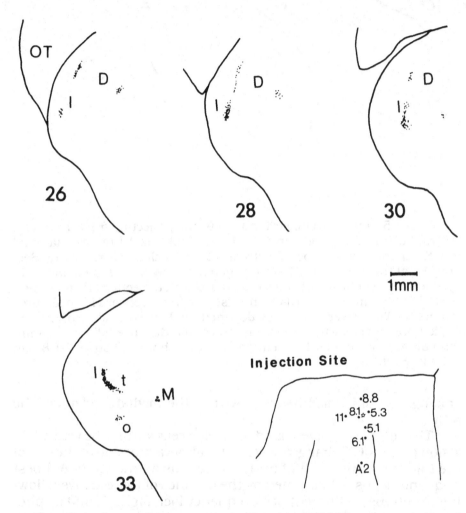

FIG. 2.4. Array of neurons projecting to a 7.5 kHz representational locus (0.1 µl HRP injection, at the site illustrated at lower right). Note the broken (two-column) array in the lateral part of the ventral division (Ss. 26–28) and the restriction of labeled neurons within a very narrow MGB band. Abbreviations and descriptions as in Fig. 2.3. OT, optic tract. Section 26 is most rostral.

from edge to edge, folded medialward into the transitional part of the lateral division, then, further posteriorly, folded back outward to form a second usually thinner band bisecting the ovoidal part of the ventral nucleus. A topographically separable cell column is seen within the deep part of the dorsal nucleus, extending rostrally into the lateral part of the posterior group. Finally, there is a distinct

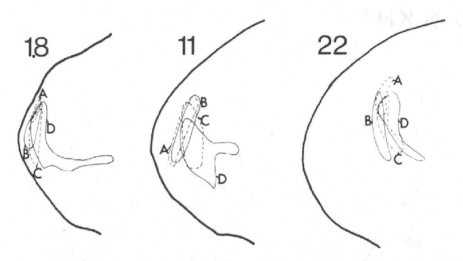

FIG. 2.5. Lines circumscribe all neurons projecting from the lateral ventral MGB at four levels, to 1.8-, 11- and 22-kHz A I representational loci. Section A is at the rostral extreme of the labeled cell array in IV. Section D is at the most rostral level at which the array extends medialward to appose labeled neurons within the medial division. Sections B and C were ⅓rd of the distance and ⅔rds of the distance from A toward D, in all three examples. The lower frequency-destined thalamocortical neurons extended over a somewhat shorter rostrocaudal distance (about 1.2 mm) than did the two higher frequency arrays (which were about 1.5-1.8 mm long) (see ref. 8).

small group of labeled neurons within the medial nucleus of the MGB.

The complexly folded band of labeled cells within the ventral division of the MGB changes systematically as a function of the site of the injection. This shift of array location as a function of A I best frequency locus is illustrated by the outlines of three arrays following injections at different best frequency loci (Fig. 2.5) and by photomicrographs of a double injection (two-frequency) case (Fig. 2.6). Successively lower frequency projection arrays were lateral to and folded within (see Fig. 2.6) the higher best-frequency projection arrays. The lower the best frequency, the shorter the anterioposterior dimension of the array, and the narrower and smaller the proportional size of the ovoidal sector of the array.

The sheet of labeled neurons within the pars ovoideus of the ventral nucleus is flat and not coiled, as might have been expected given the description of a "coiled" cell and axon orientation in this region (20, 21). The arrays in the lateral and transitional sectors of the MGB parallel the axes of Morest's defined cellular and fiber laminae.

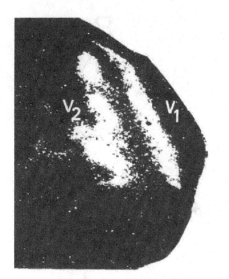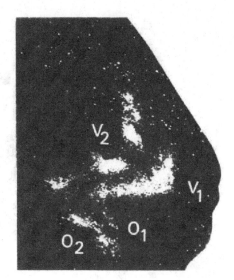

FIG. 2.6. Double array labeled by two small injections into 4.5 and 14 kHz representational loci in A I in an adult cat. The section at the left is from near the middle of the arrays in the lateral part of ventral MGB; the section at the right is from the caudal third of the arrays. The injections in this cat were combined HRP-radioactive leucine cocktails. Photomicrographs are autoradiographs; the HRP-labeled cell distributions very closely paralleled these two corticothalamic arrays (i.e., they closely overlaid projecting neurons; see Colwell and Merzenich, 8). Note the broken distribution of the label for the higher frequency injection (V_2-O_2) with 5 or 6 clear clusters of grains (columns, in three dimensions) evident in the caudal aspect of the projection.

The location of cell columns in the deep dorsal and medial divisions also shifted systematically as a function of the cortical best frequency locus of the injection site. There appears to be a reversal in best-frequency representational sequence across the boundaries between the lateral part of the ventral nucleus and the deep part of the dorsal nucleus, with highest frequencies represented along their mutual border. Another apparent reversal was evident along the border of the lateral transitional and ovoidal parts of the ventral division, with lowest frequencies represented on this border [see Colwell and Merzenich (8) for further details].

3.2. Banding of the MGB Ventral Division Projection

With injections that spread across the cortex more than about a millimeter, a *continuous*, folded sheet of projecting neurons was almost invariably observed within the ventral division. However,

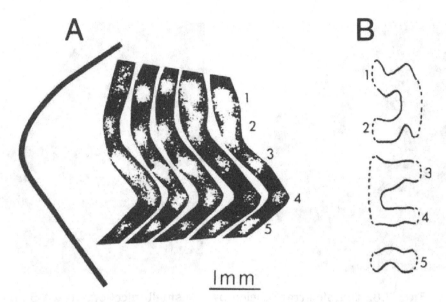

FIG. 2.7. Banding of the corticothalamic array, illustrated by example with another small-injection case into higher frequency A I. In the caudal aspect of the array, five clear bands of grains are evident, extending over about a 1 mm region. The broken label from five adjacent sections through this zone are shown apposed in the reconstruction at the left (A). This sector of the corticothalamic array is reconstructed to scale (from a lateral view) in B. These stubby bands were evident in all smaller injection higher-frequency A I representational site studies, in both corticothalamic terminal and thalamocortical neuronal arrays (see Figs. 2.3, 2.4, 2.6 for other examples). With larger injections, corticothalamic and thalamic arrays in the lateral–transitional parts of the ventral nucleus were usually continuous. Adapted from Andersen (1).

with smaller injections (especially at higher best frequency cortical representational loci), *discontinuous, banded* arrays were commonly recorded. An example is illustrated in Fig. 2.7. It is possible that these banded arrays of neurons (and corticothalamic terminals) represent a segregation of thalamic projections to the "binaural bands" of A I. That is, "binaural bands" might also exist within the ventral division of the MGB. This banding in the ventral division is further manifested by results of tracer injection studies in the ICC, in which restricted injections resulted in banded terminal arrays in the efferent projection to the MGB (2). With small cortical injections introduced at low frequency representational loci, again, the projecting MGB arrays were not discontinuous.

3.3. Descending Corticothalamic Projections from A I to the MGB.

There is a remarkably detailed reciprocal projection from the neurons of each A I locus back into the complex arrays of neurons in the MGB projecting to those loci (8). In fact, given problems in correlating material developed by the two histochemical methods, the pattern details of the two projections are very remarkably similar.

3.4. Interconnections of MGB Nuclei with Other Cortical Fields

Extensive studies of connections of the MGB with another large topographic auditory cortical field in the cat, the anterior field (AAF; see Fig. 2.1), have revealed that it derives its input from the same four principal MGB (IV, OV, M, D) sources as A I. The projection arrays of these two cortical fields also have similar geometric patterns. The principal difference is in the relative numbers of neurons in each subdivision of the MGB that project to each field. AAF receives its strongest projection from the deep dorsal division. (1, 3).

The connections of A II are primarily with nuclei that do not project to A I or AAF (the caudal aspect of the dorsal division, the ventral lateral nucleus and the medial division) (1, 3). These data and studies by other investigators are consistent with the interpretation that there are two largely parallel auditory projection pathways (1, 3; also see 6, 9), projecting to cochleotopically and noncochleotopically organized regions of auditory cortex in the cat.

4. Conclusions

The results of these studies are basically consistent with those of other investigators (esp. 10, 22, 23, 25, 26, 29) using techniques not involving false label through damage to fibers of passage, i.e., with results of studies, in which injections or lesions were in A I. Differences in interpretation of the MGB sources of neurons projecting to different cortical fields are believed to result from differences in the definition of (or in failure to define) the actual locations and boundaries of auditory cortical fields in those other studies.

The locations and boundaries of these relatively small cortical fields are inconstant (17, 18), largely because the dorsomedial termination zones of the anterior and posterior ectosylvian sulci are highly variable. Thus, physiological or cytoarchitectonic definition of injection or lesion sites (the latter is very difficult) is requisite for straightforward interpretation of such studies.

From these experimental data, the following basic conclusions about the organization of interconnections of the auditory cortex and thalamus can be drawn:

1. Complex arrays of neurons extending across four MGB subdivisions project to restricted A I loci. There is a remarkable convergence in the MGB-A I projection, from complex sheets and columns and clusters of MGB neurons to small A I sites.

2. These tracing studies provide evidence for the existence of repeating subunits in cortex, since injections in which tracer spreads over only a fraction of A I result in a continuous sheet of labeled neurons that bisect the nuclei of the MGB (i.e., extends across the nuclei from edge to edge). Thus, the *same* MGB neuronal figure must project to different (repeating) sectors along A I isorepresentational lines. This conclusion is also supported by the observation that a second injection along an isofrequency line does not add dimensionally to the projection array. We might hypothesize that at higher frequency A I representational loci, this repeating subunit is an adjacent pair of EE and EI bands.

3. Results with smaller injections indicate: (a) There is a banded structural organization of the ventral nucleus of the MGB, with the bands oriented orthogonal to isofrequency contours; and (b) that a given A I binaural band receives input from a *series* of MGB ventral nucleus bands. Again, this constitutes a remarkably complex pattern of convergence.

4. MGB "bands" appear to be established via a banded segregation of input from different response-specific sectors of the ICC to the MGB.

5. There is a remarkable reciprocity of connections between A I and MGB subdivisions.

6. MGB interconnections with another topographic cortical field in the cat, the anterior auditory field, are just as complex as with A I. The same four principal nuclear subdivisions (1V, OV, M, D) are reciprocally interconnected with this field.

The basic information processing significance of these complex connections is still unknown and they constitute a great practical problem for auditory neuroscientists. The detailed relationship with (and function of) A I (and MGB) functional subfields, their brain stem and ICC and MGB sources and the sources of different cortical band-specific inputs are all under intensive investigation. These studies are leading to an increasing understanding of the functional organization of the higher levels of this very complexly organized information handling system.

Acknowledgments

The authors would like to thank Tina Guerin, Joe Molinari and Randall J. Nelson for assistance in preparation of this manuscript. Work described herein was conducted with the support of NIH Grant NS-10414, the Coleman Fund and Hearing Research Inc.

References

1. ANDERSEN, R. A. *Patterns of Connectivity of the Auditory Forebrain of the Cat.* Thesis, University of California, San Francisco, CA, 1979.
2. ANDERSEN, R. A., ROTH, AITKIN, L. M. AND MERZENICH, M. M. The efferent projections of the central nucleus and the pericentral nucleus of the inferior colliculus in the cat. *J. Comp. Neurol.,* 194–662, 1980.
3. ANDERSEN, R. A., SNYDER, R. L., AND MERZENICH, M. M. Thalamo-cortical and corticothalamic connections of AI, AII and AAF in the cat. *J. Comp. Neurol.* 194:663–701, 1980.
4. ANDERSEN, R. A., SNYDER, R. L., AND MERZENICH, M. M. The topographic organization of corticocollicular projections from physiologically identified loci in the AI, AII, and anterior auditory cortical fields of the cat. *J. Comp. Neurol.,* 191: 479–494, 1980.
5. BRUGGE, J. F., AND IMIG, T. J. Some relationships of binaural response patterns of single neurons of cortical columns and interhemispheric connections of auditory area AI of cat cerebral cortex. In: *Evoked Electrical Activity in the Auditory Nervous System,* edited by R. G. NAUNTON, AND C. FERNÁNDEZ. New York: Academic Press, 1978, pp. 487–503.
6. CASSEDAY, J. H., DIAMOND, I. T., AND HARTING, J. K. Auditory pathways to the cortex in *Tupaia glis. J. Comp. Neurol.,* 166: 303–340, 1976.
7. COLWELL, S. A. *Reciprocal Structure in the Medial Geniculate Body.* Thesis, University of California, San Francisco, 1977.
8. COLWELL, S. A., AND MERZENICH, M. M. Relation of corticothalamic auditory projections. *J. Comp. Neurol,* in press, 1982.
9. DIAMOND, I. T. The subdivisions of neocortex: A proposal to revise the traditional view of sensory, motor and association areas. *Prog. Psychol. Physiol. Psychol.,* 8: 1–43, 1979.
10. DIAMOND, I. T., JONES, E. G., AND POWELL, T. P. S. The projection of the auditory cortex upon the diencephalon and brainstem in the cat. *Brain Res.,* 15: 305–340, 1969.
11. IMIG, T. J., AND ADRIÁN, H. O. Binaural columns in the primary field (AI) of cat auditory cortex. *Brain Res.,* 138: 241–257, 1977.
12. IMIG, T. J., AND BRUGGE, J. F. Sources and terminations of callosal axons related to binaural and frequency maps in primary auditory cortex of the cat. *J. Comp. Neurol.,* 182: 637–660, 1978.

13. KNIGHT, P. L. Representation of the cochlea within the anterior auditory field (AAF) of the cat. *Brain Res.*, 130: 447–467, 1977.
14. MERZENICH, M. M. Some recent observations on the functional organization of the central auditory and somatosensory nervous system. In: *Proceedings of the Third International Symposium on Brain Mechanisms of Sensation*, edited by Y. KATSUKI, M. SATO, AND R. NORGREN. New York: Wiley and Sons, 1981.
15. MERZENICH, M. M., ANDERSEN, R. A., AND MIDDLEBROOKS, J. C. Functional and topographic organization of the auditory cortex. *Exptl. Brain Res., Supp.* (2): 61–75, 1979.
16. MERZENICH, M. M., AND KAAS, J. H. Principles of organization of sensory-perceptual systems in mammals. In: *Progress in Psychobiology and Physiological Psychology*, edited by J. M. SPRAGUE, AND A. N. EPSTEIN. 9: 1–42, 1980.
17. MERZENICH, M. M., KNIGHT, P. L., AND ROTH, G. L. Representation of cochlea within primary auditory cortex in the cat. *J. Neurophysiol.*, 38: 231–249, 1975.
18. MERZENICH, M. M., ROTH, R. A., KNIGHT, P. L., AND COLWELL, S. A. Some basic features of organization of the central auditory system. In: *Psychophysics and Physiology of Hearing*, edited by E. F. EVANS, AND J. P. WILSON. London: Academic Press, 1977, pp. 485–495.
19. MIDDLEBROOKS, J. C., DYKES, R. W., AND MERZENICH, M. M. Binaural response-specific bands in primary auditory cortex (AI) of the cat: Topographic organization orthogonal to isofrequency contours. *Brain Res.*, 181: 31–48, 1980.
20. MOREST, D. K. The neuronal architecture of the medial geniculate body of the cat. *J. Anat. (London)*, 98: 611–630, 1964.
21. MOREST, D. K. The laminar structure of the medial geniculate body of the cat. *J. Anat. (London)*, 99: 143–160, 1965.
22. NIIMI, K. AND MATSUOKA, H. Thalamocortical organization of the auditory cortex in the cat studied by retrograde axonal transport of horseradish peroxidase. *Adv. Anat. Embryol. Cell Biol.*, vol. 57, 1979.
23. RACZKOWSKI, D., DIAMOND, I. T., AND WINER, J. Organization of thalamo-cortical auditory system in the cat studied with horseradish peroxidase. *Brain Res.*, 101: 345–354, 1976.
24. RAMON Y CAJAL, S. *Histologie du Système Nerveux de l'Homme et des Vertébrés.* (Reprinted from the original 1909-11 edition.) Madrid, Spain: Consejo Superior de Investigationes Cientificas.
25. ROSE, J. E., AND WOOLSEY, C. N. The relations of thalamic connections, cellular structure and evocable activity in the auditory region of the cat. *J. Comp. Neurol.*, 91: 441–466, 1949.
26. ROSE, J. E., AND WOOLSEY, C. N. Cortical connections and functional organization of the thalamic auditory system of the cat. In: *Biological and Biochemical Bases of Behavior*, edited by H. F. HARLOW, AND C. N. WOOLSEY. Madison, Wisconsin: University of Wisconsin Press, 1958, pp. 127–150.

27. TUNTURI, A. R. Further afferent connections of the acoustic cortex of the dog. *Amer. J. Physiol.*, 144: 389–394, 1945.

28. TUNTURI, A. R. Physiological determination of the arrangement of the afferent connections to the middle ectosylvian area in the dog. *Amer. J. Physiol.*, 162: 489–502, 1950.

29. WINER, J. A., DIAMOND, I. T., AND RACZKOWSKI, D. Subdivisions of the auditory cortex of the cat: the retrograde transport of horseradish peroxidase to the medial geniculate body and posterior thalamic nuclei. *J. Comp. Neurol.*, 176: 387–418, 1977.

30. WOOLSEY, C. N. Organization of cortical auditory system: A review and a synthesis. In: *Neural Mechanisms of the Auditory and Vestibular Systems*, edited by G. L. RASMUSSEN, AND W. F. WINDLE. Springfield, IL.: C. C Thomas, 1960, pp. 165–180.

31. WOOLSEY, C. N. Organization of cortical auditory system. In: *Sensory Communication*, edited by W. A. ROSENBLITH. Cambridge, MA: MIT Press, 1961, pp. 235–257.

32. WOOLSEY, C. N. Electrophysiological studies on thalamocortical relations in the auditory system. In: *Unfinished Tasks in the Behavioral Sciences*, edited by A. ABRAMS, H. H. GARNER, AND J. E. P. TOMAN. Baltimore, MD: Williams and Wilkins, 1964, pp. 45–57.

33. WOOLSEY, C. N. AND WALZL, E. M. Topical projection of nerve fibers from local regions of the cochlea in the cerebral cortex of the cat. *Bull. Johns Hopkins Hosp.*, 71: 315–344, 1942.

Chapter 3

Auditory Cortical Areas in Primates

John F. Brugge

Department of Neurophysiology and Waisman Center on Mental Retardation and Human Development, University of Wisconsin, Madison, Wisconsin 53706

1. Early Studies in Monkeys, Apes and Humans

Our knowledge of auditory areas of cerebral cortex in the primate begins with the published work of David Ferrier (4, 5). Following the experiments of Fritsch and Hitzig on the frontal lobe of the dog, Ferrier found that electrical stimulation of a limited region of the temporal lobe in the monkey resulted in orienting behavior reminiscent of that seen when the animal encounters a sudden novel sound. On the basis of a large series of experiments in monkeys and other mammals, in which he explored electrically much of the exposed surface of the cerebral hemispheres, Ferrier concluded that these movements evoked by stimulation of a restricted region of the superior temporal gyrus (his area 14), or its presumed homolog in nonprimate species, are outward manifestations of the arousal of subjective auditory sensations and that this responsive area is, indeed, a center of auditory sensibility. From more recent studies in the macaque monkey, it would appear that Ferrier's electrodes did not contact the primary auditory field (A I) directly, but were within one or more of the fields that are contiguous with it on the exposed gyral surface. From these and other studies like them, Ferrier clearly recognized that sensory and motor functions are localized in cerebral cortex and that there are multiple spatial sensory and motor representations within the nervous system. He further concluded that these various cortical regions are intimately tied to one another anatomically and functionally.

The primate, including the human, occupied center stage for work on auditory cortical localization during the first half-century after Ferrier's initial publications on the subject. Campbell (2), in his studies of the cytoarchitecture of the brains of apes and humans, described a cortical area having a distinct cytoarchitecture and occupying the anterior transverse gyrus of Heschl on the superior surface of the temporal lobe. Several years earlier, Flechsig (7) had concluded that this region was the primary cortical receiving area on the basis of its relatively early development. Campbell called this region "audiosensory" cortex and suggested that it might function as an "arrival platform" for primary reception of crude auditory sensations. He also distinguished on cytoarchitectonic grounds a belt of cortex surrounding his audiosensory region that he called the "audiopsychic" cortex. He suggested that this belt region, like secondary visual areas, might function in "further elaboration and transmission of crude sensations into conscious psychic perception." Thus, not long after the turn of this century the notion of dual cortical auditory areas was established and there was already speculation on the ways in which these areas might carry out their respective roles in processing acoustic information.

Little has been added to von Economo's cytoarchitectonic description of the temporal lobe in humans (3). He recognized as the primary auditory cortical field a small region of koniocortex (his area TC), which coincides with the area described by Flechsig on developmental grounds. Surrounding this field, he recognized several additional areas that could be distinguished from one another on the basis of their cellular architecture. Von Economo emphasized that, though there are architectural differences among these presumed auditory cortical fields, the boundaries between fields are often not sharp, but rather show various degrees of transition from koniocortex to the less granular structure of surrounding areas. Later, Rose (19) would point to similar gradients in cellular structure within and between auditory cortical areas in the cat.

In a tour de force of myeloarchitectonics, Beck (1) identified some 88 subdivisions of the cortex of the dorsal surface of the superior temporal lobe in the chimpanzee. Walzl and Woolsey's (23, 24, 27, 29) evoked potential maps of the region show that a portion of this area, over and immediately surrounding Heschl's gyrus (Beck's area ttr_1), is excited by acoustic stimulation and contains a topical representation of the cochlear nerve. (See Chapter 8, this volume.)

Degeneration studies of Poliak (17) and Walker (22) in the macaque monkey shed further light on the organization of primate auditory cortex. A "nuclear or focal zone" of cortex on the supratemporal plane, richly supplied with geniculocortical afferents, was described by Poliak in Marchi material. Poliak was impressed by the regular arrangement and distribution of fibers of the auditory radiation during their course toward the cortex and suggested that this might represent a spatial array of frequencies within the primary field. He even went so far as to speculate that small auditory cortical lesions or interruptions of portions of the auditory radiations would probably produce one or several gaps in tone perception comparable to scotomata of the visual field. However, it was Walker, in a retrograde degeneration study, who demonstrated for the first time in the monkey that there is an orderly projection from the medial geniculate body to the auditory koniocortical field.

Despite this early interest in auditory cortex in primates, most of the later work on the organization of this area of the brain was carried out in the cat. Indeed, the systematic evoked potential studies of Woolsey and his colleagues, beginning with the 1942 paper of Woolsey and Walzl (26) describing in the cat the cortical map obtained by electrical stimulation of small bundles of fibers in the osseous spiral lamina, have provided a scheme of auditory cortical organization which, with slight modification, stands today (27). From these and other later studies, it is abundantly clear that the belt of cortex surrounding the primary auditory field in the cat comprises not one but several distinct auditory areas. These areas can be distinguished from one another and from A I not only on the basis of their cytoarchitecture (19), but also on the basis of their tonotopic maps and connectivity patterns (9, 10, 11, 13, 18, 20, 21). In an accompanying chapter, Imig et al. (10) summarize the results of their recent anatomical and physiological mapping studies of the cat's auditory cortex.

Walzl and Woolsey, as mentioned above, also applied their nerve stimulation methods to the study of auditory cortex in the monkey, although until now only a summary figure of their results has ever been published; this shows the location and topographic organization of A I, as well as the position of a second field they called A II (23). In another unpublished evoked potential study of the macaque monkey, Kennedy (see ref. 15) showed that low frequencies were represented rostrolaterally on the supratemporal plane and that high frequencies were localized caudomedially in agreement with nerve stimulation results.

In 1971, Woolsey reviewed evoked response data on the organization of the cortical auditory system in several primates and included in his review previously unpublished observations on the chimpanzee, owl monkey and squirrel monkey (28). Area A I was identified in each of these species; in addition, evidence was presented in the owl monkey for several organized auditory fields surrounding A I. These observations were later confirmed and extended by detailed microelectrode mapping. In another paper in this volume (6), FitzPatrick and Imig report on the different auditory cortical efferent projection patterns that characterize two of the well-mapped auditory areas in the owl monkey.

2. Microelectrode Studies of the Organization of Primate Auditory Cortex

The three species of primates in which auditory cortex has been mapped with microelectrodes are shown in Fig. 3.1, along with photographs of their respective brains. The macaque monkey, an Old World simian, is shown on the left in this figure. The brain in this animal is well fissured compared to those of the other two monkeys shown here. The New World monkey, *Aotus trivirgatus* (owl monkey), is depicted in the center of Fig. 3.1. Its brain is relatively lissencephalic, showing only well developed Sylvian and superior temporal fissures. On the left appears the Old World prosimian *Galago crassicaudatus* (greater African bushbaby; thick-tailed Galago). The rather lissencephalic brain of this animal is also shown in an outline drawing in Fig. 3.2. An intraparietal sulcus (IP) is present, as well as a sulcus rectus (R). The Sylvian (S), or lateral sulcus, is well developed in the Galago.

The microelectrode mapping technique is one that was pioneered and used extensively by Welker and his colleagues in studies of somatic sensory cerebral cortex (25). It was employed by Merzenich and Brugge (12) in their work on auditory cortical organization in the macaque monkey and later by Imig and his coworkers (8) in similar but more extensive experiments on the owl monkey. Recently, in our laboratory, studies of this kind have been extended to include the greater Galago. In these studies, the distribution of best frequencies of neurons and neuron clusters is mapped under barbiturate anesthesia. After the experiment, the brain is processed

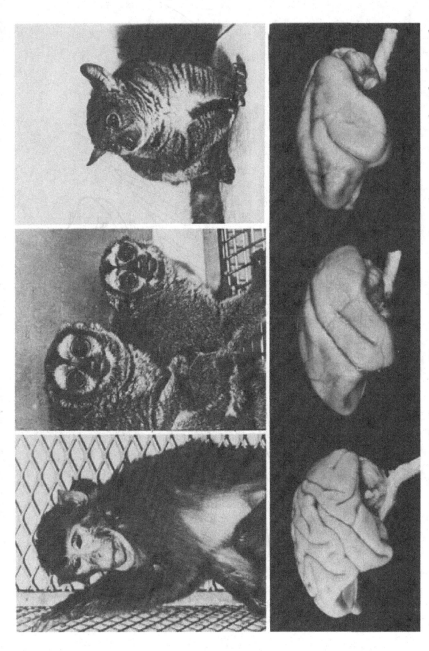

FIG. 3.1. Top row: Primates whose auditory cortical areas have been mapped with microelectrode recording methods. Left: Old World monkey, *Macaca mulatta*; middle: New World monkey, *Aotus trivirgatus*; right: prosimian, *Galago crassicaudatus*. Bottom row: lateral view of the left sides of brains of animals shown above. Photos not to scale.

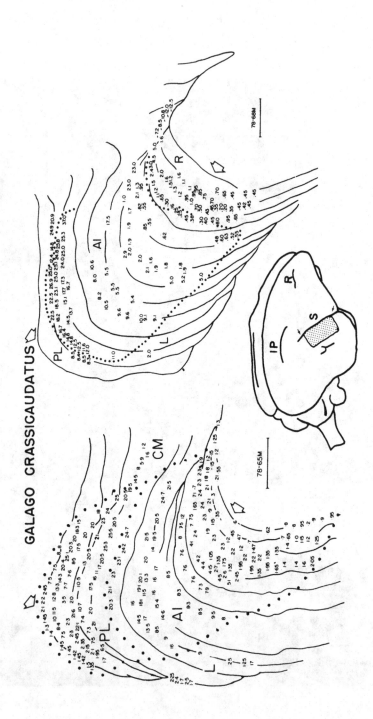

GALAGO CRASSICAUDATUS

FIG. 3.2. Tonotopic maps of auditory cortical areas of *Galago crassicaudatus* from two experiments. Contour maps constructed from serial histological sections. For clarity, not all sections are drawn here. Experiments were conducted on the right cerebral hemispheres. Lateral is to the left of each map: medial is to the right. Caudal is to the upper left, rostral to the lower right. Numbers are best frequencies of neurons or neuron clusters recorded at the respective sites. Arrows on the contour maps point to the approximate location of the lateral (Sylvian) fissure (S), shown on the brain drawing below. Rows of dots mark the boundaries of auditory fields suggested by the physiological maps. A I: primary auditory field; R: rostral field; PL: posterolateral field; CM: caudomedial field; L: lateral field.

histologically; the positions of electrode tracks are determined and a map of the distribution of best frequencies is constructed. Since the best frequency, i.e., that frequency to which an auditory neuron is most sensitive, is related to the place along the basilar membrane to which the cell is connected, it is possible by this method to determine the spatial representation of the cochlear partition within an auditory cortical area.

In the macaque monkey, area A I is found to contain a complete and orderly representation of the audible frequency spectrum, in agreement with earlier evoked potential studies. The boundaries of this tonotopic map correspond closely with the boundaries of the cytoarchitectonic field of koniocortex. In a more extensive study of primate auditory cortex, Imig and his colleagues (8) mapped the frequency representations of A I and surrounding cortical fields in the owl monkey. As in the macaque monkey, A I in *Aotus* contains a complete and orderly representation of the audible tonal spectrum with low frequencies represented rostrolaterally in the field and high frequencies caudomedially. Rather than being confined to the depths of the lateral fissure, as it is in the macaque monkey, A I in the owl monkey extends onto the crown of the superior temporal gyrus. It is also coextensive with auditory koniocortex.

In addition to firmly establishing the boundaries and frequency organization of A I, Merzenich and Brugge (12) identified possibly four other auditory fields in the macaque monkey that could be distinguished from A I both on architectonic and physiologic grounds. None of them was mapped completely, however, although in some cases the data were complete enough to suggest a tonotopic organization within the field. One of these areas, called the rostrolateral field (RL), joins the rostral border of A I. The high-to-low frequency sequence of best frequencies that is recorded as the electrode is systematically moved caudorostrally in A I reverses in this border region. In these experiments, best frequencies above about 8 kHz were not encountered in area RL, hence a complete tonotopic map of this field has not been obtained in the macaque monkey. Independently, Pandya and Sanides (16) identified this area on the basis of its cellular architecture alone. In later mapping experiments in the owl monkey, Imig and his colleagues fully explored the region immediately rostral to A I and found there a complete and organized tonal representation (8). They refer to this area simply as the rostral field (R). Its cytoarchitecture closely resembles that of area RL in the macaque monkey and we may consider the two regions to be homologous. Caudomedial to A I is an area whose cytoarchitecture is distinct from that of A I. It is referred to as the caudomedial field (CM). In tonotopic maps of the area in both macaque and owl monkeys, the border region is usually identified physiologically by the apposition of the high frequency region of A I and low frequency representation of CM. A complete map of CM has not been obtained in any primate species.

In addition to identifying areas A I, R and CM, Merzenich and Brugge (12) and Imig et al. (8) recorded from responsive neurons caudal, lateral and medial to A I, within the inferior limiting sulcus adjacent to the insula. Auditory cells were also encountered rostral to field R. None of these areas was mapped completely in either the macaque or owl monkey.

Recently, our group at Wisconsin has carried out a series of single unit mapping experiments in auditory cortex of *Galago crassicaudatus*. From the point of view of morphology, the Galagos are placed with Lorises in the family *Lorisidae*, which has representatives in Asia and Africa (14). In primate evolution, this group can be viewed as occupying a rung below the New and Old World monkeys, keeping in mind of course that any simple graded series of living primates represents the end products of many diverging lines of evolution. Thus, it was of great interest to us to explore the extent to which the organization of a sensory cortex in a prosimian primate resembles that of Old and New World monkeys.

In Fig. 3.2, the stippled area on the lateral surface of the superior temporal gyrus represents the exposed region that is responsive to acoustic stimulation. Auditory responsive cortex also extends into the lateral fissure and occupies much of the superior temporal plane. The results of two detailed mapping experiments are presented in Fig. 3.2. In each case, a contour map of the superior temporal gyrus, reconstructed from serial histological sections, is shown. The maps are oriented so that the section at the upper left is most posterior and that at the lower right is more anterior. The medial edge of the gyrus is at the right and the lateral edge is at the left. Numbers indicate the best frequencies, in kHz, for cells recorded at the corresponding points on the map. Black dots indicate the borders of auditory fields based on the physiological maps.

In experiment 78-65M, shown on the left in Fig. 3.2, two nearly complete representations of the basilar membrane are evident. The larger of the two representations is located more rostrally on the gyrus. We take this one to be the primary field, A I. Here low frequencies are rostrolaterally. The frequency increases in an orderly manner as the exploring electrode progresses caudomedially. The highest best frequencies recorded appear within the caudomedial portion of this field. Within any single cross section through the superior temporal gyrus, there is a tendency for best frequencies to increase as the electrode passes from the lateral cortical surface medially through the superior temporal plane. The location and tonotopy of this field differ in no obvious or systematic way from those of A I in macaque and owl monkeys. A second, smaller area for frequency representation appears contiguous with the posterior border of A I. This area is termed the posterolateral field (PL). Here, lowest best frequencies are located laterally and highest best frequencies appear medially, deep on the superior temporal plane, adjacent to the high frequency region in A I. In fact, from the physiological map alone, it is difficult to place with certainty the actual border between A I and PL in regions where best frequencies in the two fields are so similar. An auditory field in this location was identified by Imig et al. (8) in the owl monkey, but in those experiments there were too few data points to map either the boundaries of the field or its tonotopic organization. In the macaque monkey, Merzenich and Brugge (12) also found neurons in the area posterior to A I that responded to auditory stimulation. However, these results were not complete enough to identify the responsive cells as belonging to a separate auditory field. Responsive neurons are encountered lateral to A I and PL. This area is termed the lateral field (L). Low-best frequencies predominated in this experiment and no tonotopy is evident. Cells responsive to sound were also encountered medial to A I and PL in an area termed the caudomedial field (CM). In one penetration shown here, which traversed both PL and CM, the increasing sequence of best frequency seen as the electrode passed medially through PL began to decline (from about 20 to 1.2 kHz) as the electrode entered and continued medially through CM. Both CM and L were identified in the macaque and owl monkeys, although in these cases, as with the Galago, the boundaries and organizations of the fields were not fully mapped.

A second mapping experiment (78-68M) is shown to the right in Fig. 3.2. A I is organized as previously described. Areas PL and CM were not mapped in this experiment, although the 8.3–8.5 kHz points seen just posterior to A I may belong to PL. In this case, a reversal in the best frequency sequence occurs at the rostral border of A I. As the electrode moves anteriorly and medially out of A I, best frequencies, which have been quite low in this border region, begin to increase again, reaching 12.5 kHz within the inferior limiting sulcus. Thus, there is an indication that another tonotopically organized field lies rostral to A I in the Galago. This area may be the homolog of the one described earlier in the macaque and owl monkeys as the rostral field (R). We were unable in these experiments to explore the more anterior regions of the gyrus. Thus, it is not known whether this field contains a representation of the upper octave or two of the tonal spectrum.

Acknowledgments

Experiments on *Galago crassicaudatus* were carried out in collaboration with Kathleen FitzPatrick (Merrimac College), Carol Welt (Central Wisconsin Center), Thomas O'Connor (Dept. of Neurophysiology, Univ. of Wisconsin) and Barbara Sprague (Dept. of Communicative Disorders, Univ. of Wisconsin).

This work was supported by NIH grants HDO3352, NS12732 and NS5646.

References

1. BECK, E. Der myeloarchitektonishe Bau des in den Sylvischen Furche gelegenen Teiles des Schlafenlappens beim Chimpansen (*Troglodytes niger*) *J. Psychol. Neurol., Leipzig*, 38: 309–420, 1929.

2. CAMPBELL, A. W. *Histological Studies on the Localisation of Cerebral Function*. Cambridge: Cambridge Univ. Press, 1905.

3. ECONOMO, C. VON. *The Cytoarchitectonics of the Human Cerebral Cortex*, translated by S. PARKER. London: Oxford University Press, 1929.

4. FERRIER, D. *The Functions of the Brain*. London: Smith, Elder 1876.

5. FERRIER, D. *The Croonian Lectures on Cerebral Localisation. Lecture IV*. London: Smith, Elder, 1890.

6. FITZPATRICK, K. A. Organization of connections of primate auditory cortex. Chap. 4, this volume.

7. FLECHSIG, P. E. *Gehirn und Seele*. Leipzig: Verlag von Veit, 1896.

8. IMIG, T. J., RUGGERO, M. A., KITZES, L. M., JAVEL, E., AND BRUGGE, J. F. Organization of auditory cortex in the owl monkey (*Aotus trivirgatus*). *J. Comp. Neurol.*, 171: 111–128, 1977.

9. IMIG, T. J., AND REALE, R. A. Patterns of corticocortical connections related to tonotopic maps in cat auditory cortex. *J. Comp. Neurol.*, 192: 293–332, 1980.

10. IMIG, T. J., REALE, R. A., AND BRUGGE, J. F. Patterns of corticocortical projections related to physiological maps of cat's auditory cortex. Chapter 1, this volume.

11. KNIGHT, P. A. Representation of the cochlea within the anterior auditory field (AAF) of the cat. *Brain Res.*, 130: 447–467, 1977.

12. MERZENICH, M. M., AND BRUGGE, J. F. Representation of the cochlear partition on the superior temporal plane of the macaque monkey. *Brain Res.*, 50: 275–296, 1973.

13. MERZENICH, M. M., KNIGHT, P. L., AND ROTH, G. L. Representation of cochlea within primary auditory cortex in the cat. *J. Neurophysiol.*, 38: 231–249, 1975.

14. NAPIER, J. R., AND NAPIER, P. H. *A Handbook of Living Primates.* London Academic Press, 1967.

15. NEFF, W. D. Neural mechanisms of auditory discrimination. In: *Sensory Communication*, edited by W. A. ROSENBLITH, Cambridge: MIT Press, 1961, pp. 259–278.

16. PANDYA, D. N., AND SANIDES, F. Architectonic parcellation of the temporal operculum in rhesus monkey and its projection pattern. *Z. Anat. Entwickl.*, 139: 127–161, 1973.

17. POLIAK, S. *The Main Afferent Fiber Systems of the Cerebral Cortex in Primates.* Berkeley: Univ. of California, 1932.

18. REALE, R. A., AND IMIG, T. J. Tonotopic organization in auditory cortex of the cat. *J. Comp. Neurol.*, 192: 265–292, 1980.

19. ROSE, J. E. The cellular structure of the auditory region of the cat. *J. Comp. Neurol.*, 91: 409–440, 1949.

20. ROSE, J. E., AND WOOLSEY, C. N. The relations of thalamic connections, cellular structure and evocable electrical activity in the auditory region of the cat. *J. Comp. Neurol.*, 91: 441–466, 1949.

21. ROSE, J. E., AND WOOLSEY, C. N. Cortical connections and functional organization of the thalamic auditory system of the cat. In: *Biological and Biochemical Bases of Behavior*, edited by H. F. HARLOW, AND C. N. WOOLSEY. Madison: Univ. Wisconsin Press, 1958, pp. 127–150.

22. WALKER, A. E. *The Primate Thalamus.* Chicago: University of Chicago Press, 1938.

23. WALZL, E. M. Representation of the cochlea in the cerebral cortex. *Laryngoscope*, 57: 778–787, 1947.

24. WALZL, E. M., AND WOOLSEY, C. N. Cortical auditory areas of the monkey as determined by electrical excitation of nerve fibers in osseous spiral lamina and by click stimulation. *Fed. Proc.*, 2: 52, 1943.

25. WELKER, W. Mapping the brain. Historical trends in functional localization. *Brain, Behav. Evol.*, 13: 327–343, 1976.

26. WOOLSEY, C. N., AND WALZL, E. M. Topical projection of nerve fibers from local regions of the cochlea to the cerebral cortex of the cat. *Bull., Johns Hopkins Hosp.*, 71: 315–344, 1942.

27. WOOLSEY, C. N. Organization of cortical auditory system: A review and synthesis. In: *Neural Mechanisms of the Auditory and Vestibular Systems*, edited by G. L. RASMUSSEN, AND W. WINDLE. Springfield: Charles C Thomas, 1960, pp. 165–180.

28. WOOLSEY, C. N. Tonotopic organization of the auditory cortex. In: *Physiology of the Auditory System*, edited by M. B. SACHS. Baltimore: National Educational Consultants, 1971, pp. 271–282.

29. WOOLSEY, C. N., AND WALZL, E. M. Topical projection of the cochlea to the cerebral cortex of the monkey. *Amer. J. Med. Sci.*, 207: 685–686, 1944.

Note Added in Proof: Recently (GALABURDA, A., and SANIDES, F., Cytoarchitectonic organization of the human auditory cortex, *J. Comp. Neurol.*, 190: 597–610, 1980), described the cytoarchitecture of the human auditory cortex in such a way as to make comparisons with the monkey.

Chapter 4

Organization of Auditory Connections

The Primate Auditory Cortex

Kathleen A. FitzPatrick and Thomas J. Imig

Department of Neurophysiology, Waisman Center on Mental Retardation and Human Development, University of Wisconsin, Madison, Wisconsin

1. Introduction

Both evoked potential (91, 97) and microelectrode (55) mapping experiments suggest that primate auditory cortex contains multiple representations of the basilar membrane. Auditory cortex of the owl monkey can be divided into five fields using electrophysiological maps and cytoarchitectonic criteria (37). The positions of these auditory cortical fields on the superior temporal gyrus are depicted in Fig. 4.1. Both the primary field (A I) and the field rostral to it (R) contain complete frequency representations and differ from each

71

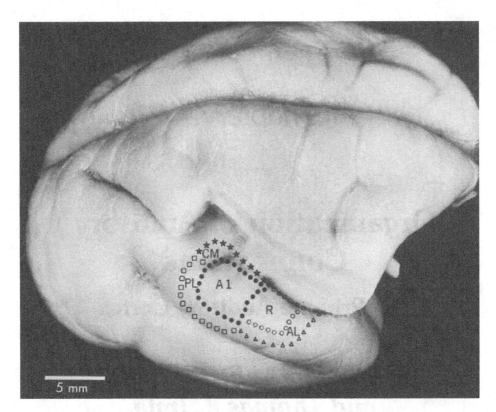

FIG. 4.1. Photograph of the right hemisphere of the brain of the owl monkey. Portions of the parietal operculum have been removed to show the superior temporal plane; the positions of the five auditory fields (37) are indicated. A I, primary field; R, rostral field; CM, caudomedial field; PL, posterolateral field; AL, anterolateral field. From (26).

other cytoarchitechonically. Surrounding these two fields are several secondary areas, the anterolateral (AL), posterolateral (PL) and caudomedial (CM) fields. The rostromedial field (RM) is not pictured in Fig. 4.1, but occupies the lateral bank of the inferior limiting sulcus. Each of these secondary fields is cytoarchitectonically distinct from the others and cells within each of them respond to acoustic stimulation.

It has been known for some time that neurons in auditory cortex in the primate project upon the medial geniculate body and inferior colliculus (29, 40, 48, 57, 79, 85, 92), as well as upon the adjacent auditory cortical areas of the ipsilateral hemisphere and contralateral hemispheres (3, 4, 17, 18, 23, 30, 34, 38, 39, 45, 53, 70–73, 84). Nevertheless, we have little information regarding similarities and differences in the projections of individual cortical fields. In this report, we describe the projections of fields A I and R upon midbrain, thalamic and cortical areas in the owl monkey (26, 27). Ultimately, knowledge of connectivity patterns may give us insight into the role that these cortical areas play in hearing.

2. Cytoarchitecture of the Medial Geniculate Body

The medial geniculate body is an architectonically complex structure and its various parts do not receive equivalent projections from fields A I and R. In accord with the division first proposed by Cajal (75) and followed most recently by Berman and Jones (5), we have divided the medial geniculate into principal, dorsal and magnocellular nuclei. We have further subdivided the dorsal nucleus into anterior and posterior divisions.

Figure 4.2 shows the cytoarchitecture of the medial geniculate body in a series of six Nissl-stained sections. The brain was sectioned in an oblique plane oriented perpendicular to the lateral fissure, thus cutting the medial geniculate in a plane parallel to the long axis of the brain stem, i.e., horizontally. Each section is oriented with anterior at the top and medial at the left. Figure 4.2A is the most dorsal section.

At the level of Fig. 4.2A, most of the medial geniculate is comprised of the posterior division of the dorsal nucleus (MGDP). Here MGDP is composed of small widely spaced cells. At the most posterior edge of the geniculate is a thin cell-sparse band, the posterior marginal zone (PMZ). Anteromedially, the large darkly stained cells of the supragemiculate—limitans complex (SG) of the posterior thalamus are evident.

In Fig. 4.2B, both divisions of the dorsal nucleus are present. The posterior division constitutes the posterior two-thirds of the medial geniculate. Anteriorly, the anterior division of the dorsal nucleus (MGDA) appears as a comma-shaped mass of medium-sized, darkly staining, densely packed cells. The anterior division is clearly separated from the posterior division by a cell-sparse band.

At the level of Fig. 4.2C, the characteristic comma shape of MGDA is still evident, but the appearance of the posterior division has changed. This characteristically cell-sparse region is now interrupted by islands of slightly larger densely packed cells. At this level, the magnocellular nucleus (MGM) is located medial to the anterior and posterior divisions of the dorsal nucleus. Cells of the magnocellular nucleus are darkly stained, loosely packed and heterogeneous in size.

At the level of Fig. 4.2D, there is no distinct border between

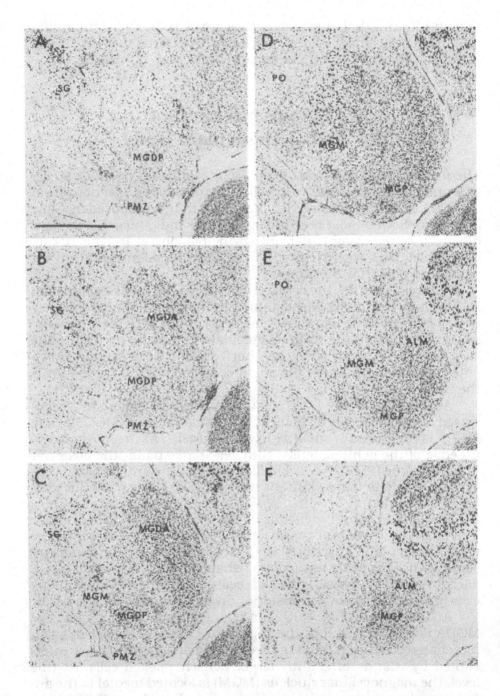

FIG. 4.2. Cytoarchitecture of the medial geniculate body of the owl
monkey. The brain was sectioned in a plane perpendicular to the lateral
fissure, thus cutting the medial geniculate nearly parallel to the long axis
of the brain stem or horizontally. Six sections, spaced at 350-μ intervals

MGM and the cell groups of the posterior thalamus (Po). Within the medial half of the magnocellular nucleus is a distinct group of large, darkly staining cells that gives the magnocellular nucleus its name. The magnocellular region is particularly evident in this plane of section in the owl monkey. The caudal and lateral part of the medial geniculate in Fig. 4.2D is occupied by the principal nucleus (MGP). It is a region of closely packed medium to small, darkly staining cells. At the anterolateral edge of the medial geniculate, there is a distinct group of cells that are widely spaced. This group will be referred to as the anterolateral margin (ALM).

In Fig. 4.2E, the appearance of the anterolateral margin and principal nucleus is similar to their appearance in Fig. 4.2D, but the large dark cells of the magnocellular nucleus are no longer obvious.

Figure 4.2F is the most ventral section through the medial geniculate. Only the MGP and ALM are present. Many cells of the ALM at this level are spindle shaped and are oriented parallel to the anterolateral margin of the medial geniculate.

Our principal nucleus corresponds to the ventral division of Burton and Jones (12); the anterior and posterior divisions of the dorsal nucleus in the owl monkey correspond to their anterodorsal and posterodorsal divisions of the principal nucleus. The magnocellular nucleus in the owl monkey corresponds to that described in the rhesus and squirrel monkey by Burton and Jones (12). Others have parcellated the medial geniculate of primate into parvocellular and magnocellular divisions [Walker and Fulton (90) in chimpanzee; Walker (89) and Mesulam and Pandya (56) in rhesus]. The parvocellular division corresponds to the principal and dorsal nuclei in the owl monkey, whereas the magnocellular division corresponds to the large-celled portion of the magnocellular nucleus in the owl monkey.

3. Projections to the Medial Geniculate

Fibers arising from cells in both A I and R leave the cortex in the auditory radiations and run in bundles through the putamen, finally reaching the medial geniculate body. The projection of A I and R upon the thalamus in the owl monkey is exclusively ipsilateral.

FIG. 4.2. (continued)
through the geniculate, are represented. Section A is the most dorsal section and F is the most ventral. In each figure sections are oriented with anterior at the top and medial at the left. Bar indicates 1 mm. Thionin stain. From (26).

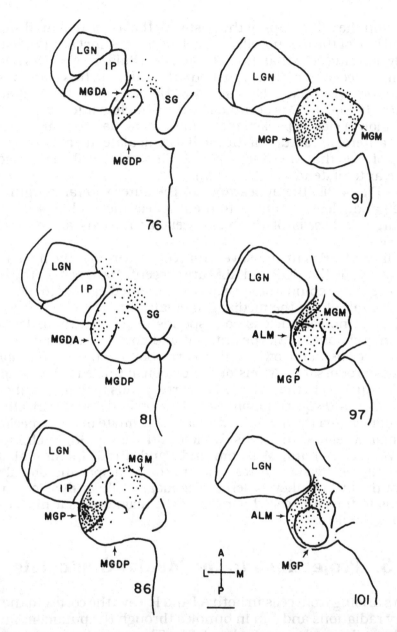

FIG. 4.3. Projection of the primary field upon the medial geniculate body. Line drawing of six nearly horizontal sections through the MGB. The upper left section is the most dorsal section and the lower right is most ventral. Each is oriented with anterior at the top and medial at the right. Bar indicates 1 mm. Density of dots indicates relative density of autoradiographic label in these sections following an injection that was confined to A I (75-158M). From (26).

After an injection in the primary field, the heaviest labeling is found in the principal nucleus of the medial geniculate; lighter label is present in MGM and ALM. Figure 4.3 shows an example of one experiment. Heaviest label is seen in the middle (Section 91 or S91) and middorsal (S86) parts of the principal nucleus. Lighter label is visible in MGM (S91, 86, 81) and in ALM (S97) and a few scattered grains are present in MGDA. It is not possible to determine whether labeling of ALM is terminal, since fibers entering the geniculate traverse the region. The pattern of labeling in MGP and MGM is illustrated in light and dark field photomicrographs in Fig. 4.4A. Other injections confined to A I (3 cases) show similar results.

Following injections confined to the rostral field, heaviest labeling is seen in the principal nucleus and the posterior division of the dorsal nucleus. The magnocellular nucleus is labeled to a lesser extent. Figure 4.5 shows the distribution of silver grains in the MGM following an injection in R. Heavy label is evident in a restricted portion of MGP (S121 and 116) and MGDP (S101, 96, 91). Moderate labeling is evident in MGM (S116, 111). The appearance of labeling in MGP, MGM and MGDP in this experiment is illustrated in the photomicrographs in Figs. 4.6A and 4.6B. In summary, MGP receives heavy projections from both A I and R. MGM receives relatively less input from R than from A I. MGDP receives input only from R and MGDA does not appear to receive a significant projection from either A I or R.

Cells in the superior temporal gyrus of the rhesus (19, 48, 57, 85, 92) and squirrel monkey (29) project to both the parvocellular and magnocellular divisions of the medial geniculate. Detailed comparison of our findings with previous studies is made difficult, however, because of insufficient documentation of the cytoarchitectonic areas which were ablated, and the use of different species of primates.

The thalamocortical connections in primates have been studied in greater detail than corticothalamic connections. Lesions involving, but not limited to, temporal koniocortex, produce retrograde degeneration in the medial geniculate (49, 79, 88). Bucy and Klüver (11) and Akert et al. (1) removed the area of superior temporal gyrus rostral to classical auditory koniocortex in the rhesus monkey and observed marked retrograde degeneration in the posterior pole of the medial geniculate. Mesulam and Pandya (56) made lesions in various parts of the medial geniculate and examined the anterograde degeneration in temporal cortex of the rhesus monkey. The anterior part of the parvocellular division of the medial geniculate sends it axons to A I, whereas the posterior parvocellular geniculate projects upon the parakoniocortex rostral to A I. This area cor-

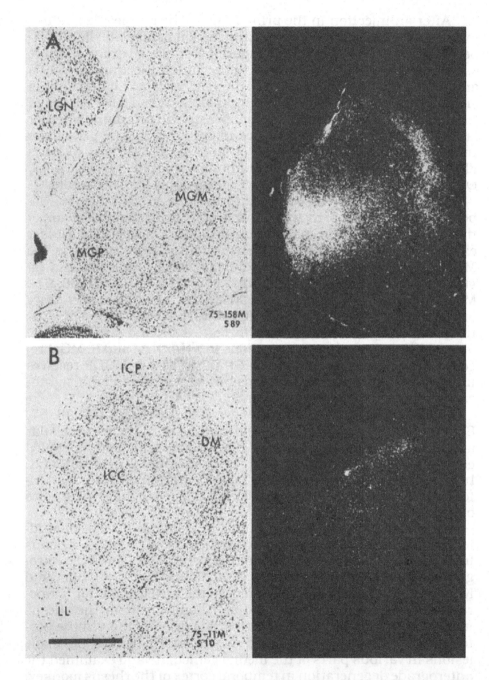

FIG. 4.4. Projection of the primary field upon the medial geniculate body and inferior colliculus. A, bright field (left) and dark field (right) photomicrographs of labeling in a nearly horizontal section through the medial geniculate following an injection in A I in experiment 75-158M. Section is 25 μ thick and is oriented with anterior at the top and medial at the right. Cresyl violet stain. B, labeling in the inferior colliculus following an

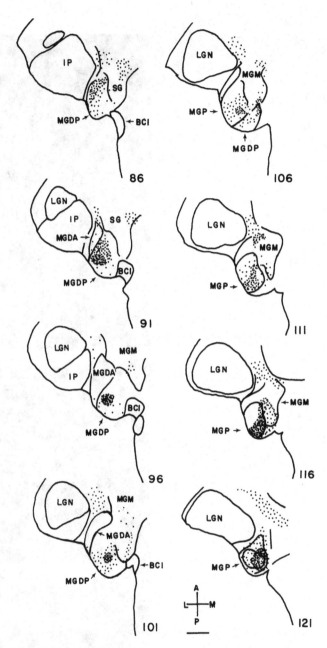

FIG. 4.5. Labeling in MGB following an injection confined to the rostral field (75-153M). See Fig. 4.3 for further details. From (26).

FIG. 4.4. (continued)
injection confined to A I (75-11M). This is a coronal section; dorsal is at the top and medial is at the right. Bar in B indicates 1 mm. in all four photomicrographs. From (26).

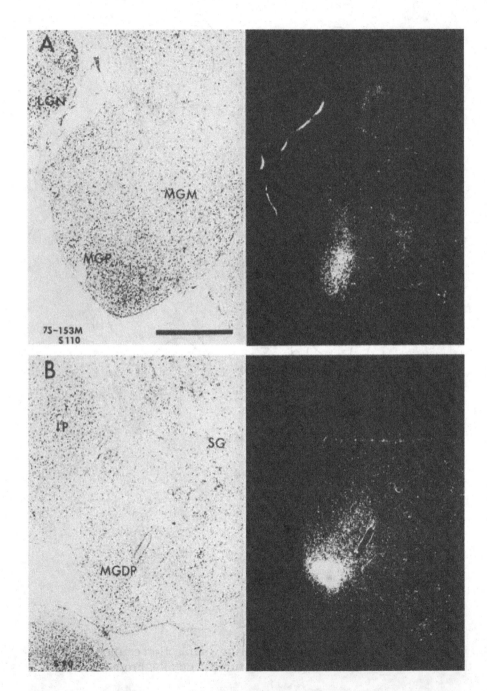

FIG. 4.6. Projection of the rostral field upon the medial geniculate body. A, labeling in a nearly horizontal section through the MGB following an R injection. Sections are oriented with anterior at the top and medial at the left. B, a more dorsal section through the same MGB. Cresyl violet stain, 25 μ section. Bar indicates 1 mm. From (26).

responds to the rostrolateral and two other rostrally located unnamed fields of Merzenich and Brugge (55) and to fields R, RM and AL in the owl monkey (37). Lesions within the magnocellular division do not appear to produce degeneration within either A I or RL. More recently, Burton and Jones (12) have used the autoradiographic tracing method to examine the projection of various portions of the posterior thalamus upon cortex. The ventral division of the principal nucleus sends its axons to A I. The small-celled portion of the posterodorsal division projects upon field RL, corresponding to R in owl monkey, and the rest of the postero-dorsal division is connected with field PL, corresponding to "A" of Merzenich and Brugge and parts of CM and RM in owl monkey. The magnocellular nucleus of the medial geniculate projects diffusely upon all the auditory fields. In the above studies, the ventral division of the principal nucleus was found to project to A I and at least part of the posterior division of the dorsal nucleus was found to project to RL. Furthermore, a projection was described to both RL and A I (and other areas) from the magnocellular nucleus. If similar thalamocortical connections obtain in the owl monkey, then it appears that A I is reciprocally connected with MGP and MGM, and R with MGM and MGDP. On the other hand, R appears to project to MGP, yet receives no projection from it.

4. Projections to Other Thalamic Nuclei

In a number of primates, the cortex of the superior temporal gyrus both projects to and receives fibers from various regions of the pulvinar (10, 16, 19, 29, 49, 50, 79, 81, 82, 87, 92). In rhesus monkey, Simpson (80) showed cell loss in medial pulvinar just above nucleus limitans after discrete lesions in an area of cortex apparently corresponding to field RL, although Burton and Jones (12) did not report a projection to RL from the medial pulvinar. Locke (51) demonstrated that the posterior part of the medial pulvinar projects to anterior temporal cortex and the medial part of the superior temporal plane. Every injection into the rostral field of the owl monkey resulted in labeling of a small region of the medial nucleus of the pulvinar just anterior to the brachium of the superior colliculus (Fig. 4.8A, S26; Fig. 4.8B, S56). No labeling was evident anywhere in the pulvinar after A I injections.

In our study, the suprageniculate nucleus, a part of the posterior complex of the thalamus, occasionally received a sparse projection from A I and R. (Fig. 4.3, S81; S76; Fig. 4.5, S91; S86). In monkeys, the suprageniculate has been shown to receive projections from the anterior temporal region and insula (40), from superior temporal gyrus (19, 29), from primary and secondary somatic cortex (44) and from spinal cord (52). Burton and Jones (12) reported that the suprageniculate–limitans complex projects to the granular insular field in rhesus and squirrel monkey. These data, along with our own, indicate that the posterior group is not a main target of output from A I or R. We cannot rule out the possibility that the sparse labeling seen in SG in our experiments may have been the result of diffusion of the amino acid from the injection site into the insula or parietal cortex.

5. Projections to the Inferior Colliculus

The cortex of the superior temporal gyrus projects to the inferior colliculus in primates (11, 57, 85, 92). Kuypers and Lawrence (48), in the rhesus monkey, and Forbes and Moskowitz (29), in the squirrel monkey, noted degeneration in the central nucleus and peripheral parts of the colliculus after lesions in the superior temporal gyrus. In the latter study, the pattern of degeneration in the colliculus appeared to depend upon the size and location of the lesion.

The cellular structure of the inferior colliculus has been described in the squirrel monkey (25). In Nissl- and Golgi-stained sections, the owl monkey inferior colliculus appears virtually identical to that of the squirrel monkey. The inferior colliculus is made up of a central nucleus surrounded dorsally and caudally by the pericentral nucleus and laterodorsally by the external nucleus. The central nucleus itself can be divided into a ventrolateral region of small spindle-shaped cells, whose somata and dendrites are arrayed in layers, and a dorsomedial region of large cells, whose dendrites do not display a laminar arrangement.

Illustrated in Fig. 4.7 are three experiments, in which injections were made into the primary field. Brains were sectioned coronally and each section is oriented with dorsal at the top and medial at the right. In each series, the upper left section is most anterior. In each case, labeling is evident in the central nucleus (ICC) of the inferior colliculus. Silver grains in the central nucleus are arrayed in a strip or band oriented in a dorsomedial to ventrolateral direction (Fig. 4.7A, S2, S5 and 11; Fig. 4.7B, S6, 11 and 16; Fig. 4.7C, S1 and 6). This orientation is roughly parallel to the orientation of isofrequency contours and dendritic laminae described in the central nucleus of the inferior colliculus of the squirrel monkey (25). These bands of label extend from the laminated region of the central nucleus into the dorsomedial region of large cells. In Fig. 4.7B, labeling is particularly heavy in the dorsomedial region (S6)

FIG. 4.7. Line drawing of labeling in the inferior colliculus in three brains following A I injections. Brains were cut coronally and in each series the section at the upper left is the most anterior and the section at lower right is most posterior. Sections are oriented with dorsal at the top and medial at the right. In each case, the injection was confined to A I (74-192M, 75-11M and 74-188M). Sections are spaced at 375 μ intervals. Bar indicates 1 mm. Density of dots indicates relative density of label. From (26).

and lighter labeling extends into the small-celled ventrolateral region of the central nucleus. This labeling is illustrated in the photomicrograph of Fig. 4.4B. Although, in a single section, these labeled regions appear as bands, they are found in several sections and thus in three dimensions they take on a disc-like form. No label is seen anywhere outside the central nucleus following injections confined to A I. In the cat (77) and tree shrew (67), primary cortex does not appear to project to the laminated portion of the central nucleus. The significance of such marked difference between the primate and other species is unknown. The pattern of the labeling seen in the inferior colliculus following injections of the rostral field differs from that seen after A I injections. After R injections, silver grains are found distributed in the cell groups surrounding the laminated portion of the central nucleus but not within it. Silver grains tend to be arranged around the edges of the laminated portion of the central nucleus. Figure 4.8 shows examples of this pattern following injections into R. Brains were sectioned in a plane perpendicular to the lateral fissure, which cuts the inferior colliculus parallel to the long axis of the brain stem or nearly horizontally. Sections are oriented with anterior at the top. Medial is at the right. In each series, the section at the upper left is the most ventral one. In Fig. 4.8A, labeling is evident in pericentral (ICP, S11, 16,

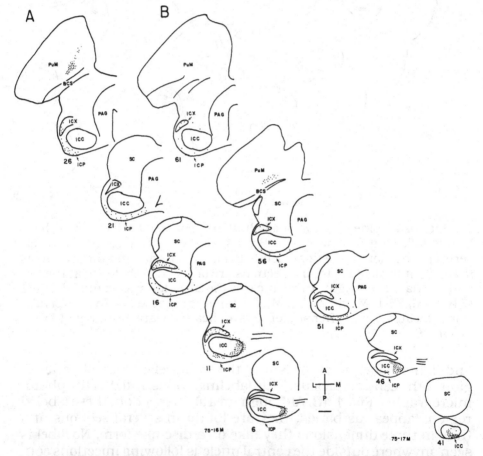

FIG. 4.8. Line drawing of labeling in the inferior colliculus and the pulvinar in two brains following injections in R. The inferior colliculi were sectioned in a nearly horizontal plane. In each series, the section at upper left is the most ventral and that at lower right is most dorsal. Sections are spaced at 375 μ intervals and are oriented with anterior at the top and medial at the right. In 75-17M, the injection was confined to R. In 75-16M, the injection was centered in R, but invaded layers V and VI of AL and perhaps a small portion of A I. Bar indicates 1 mm. From (26).

21, 26) and external nuclei (ICX, S11, 16). Grains are also seen concentrated in the dorsomedial region of the central nucleus (S6, 11). This same pattern is seen consistently after R injections (Fig. 4.8B). Thus, it appears that A I projects primarily to the dorsomedial and laminated regions of the central nucleus, while the rostral field sends its axons to the cell groups surrounding the laminated region, i.e., the external and pericentral nuclei and the dorsomedial region.

6. Cortical Projections

Before describing in detail the corticocortical projections from fields A I and R, it seems appropriate to mention a feature of the projections from both fields, which is particularly striking. In many experiments, a single injection of proline in the rostral or primary field resulted in multiple labeled columns or patches of tissue in target fields in the ipsilateral and contralateral hemispheres. Often, multiple labeled columns of tissue are visible in a single tissue section. The results of two A I injections are illustrated in Fig. 4.9 (A,B,D,E,F). Within ipsilateral RM (Fig. 4.9A), three separate patches of label are evident. Two dense columns of label can be seen within ipsilateral R (Fig. 4.9B) and are separated by a more lightly labeled area. An injection site in A I is illustrated in Fig. 4.9D and, medial to it, two separate labeled areas can be seen in CM. The same A I injection, shown in Fig. 4.9D, produces labeling of two separate areas within the contralateral A I (Fig. 4.9F). One column of tissue located more laterally is heavily labeled and a second more medial patch is lightly labeled. There is some suggestion of two patches of label in contralateral R (Fig. 4.9E) resulting from an injection in A I. The result of a rostral field injection is shown in Fig. 4.9C. Three separate columns of tissue within the ipsilateral A I are separated by areas of lighter label.

7. Ipsilateral Cortical Projections

Various investigators have demonstrated ipsilateral corticocortical connections within the temporal lobe of macaque monkeys and noted that neurons of the primary auditory field project to surrounding cortical areas (34, 70, 72, 73, 84). In the cat as well, it has been demonstrated that A I is connected with surrounding auditory fields (21, 36).

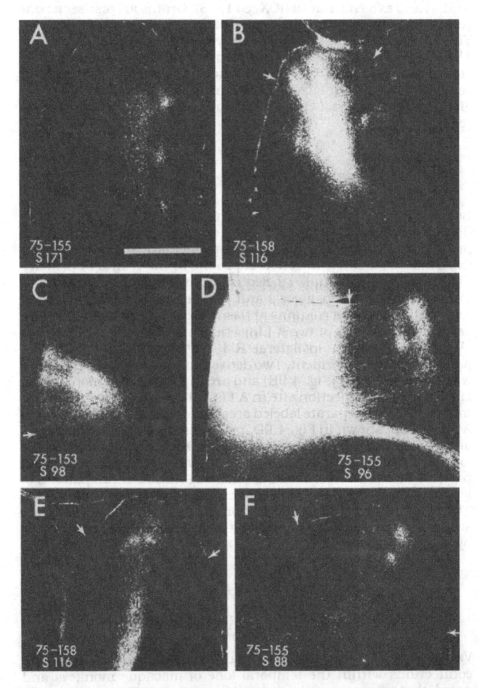

FIG. 4.9. Dark field photomicrographs of the cortical projections from the primary and rostral fields. In each, sections are cut in an oblique plane oriented perpendicular to the length of the Sylvian fissure and are oriented with dorsal at the top in each figure. In A–D, sections are oriented

Figure 4.10 (75-153M) shows the laminar distribution of ipsilateral cortical labeling resulting from an injection completely confined to the rostral field (S161 and S181). Surrounding the heavily labeled region, diffuse labeling is apparent (S181, 161, 141 and 121). Just beyond the zone of diffuse labeling in S181, increased concentrations of silver grains are apparent in AL and RM. In both fields, label is most heavily concentrated in layers III and IV.

Posterior to the injection site, the lateral part of A I contains three distinct patches of label (Fig. 4.10, S101; Fig. 4.9C), although the most ventral patch is only lightly labeled. More posteriorly (S85), these patches merge. A higher density of silver grains is found in layers I, II, V and VI than in layers III and IV (Fig. 4.11A). A labeled wedge of cortex is also seen in PL, with silver grains distributed in all layers (Fig. 4.10, S85), although layer V contains slightly less label. A surface map for this experiment appears in Fig. 4.13. These results, along with seven other cases, indicate that neurons in field R project ipsilaterally to fields A I, RM, AL and PL.

An example of ipsilateral labeling, that resulted from an injection confined to A I (75-158M), is illustrated in Fig. 4.12. In Fig. 4.12 anterior to the injection site, layer IV of RM and AL is slightly labeled (S121). The rostral field is heavily labeled with a dense accumulation of silver grains in layers III, IV, V and VI (Fig. 4.12, S121; Fig. 4.11C). Lower layer III is somewhat less heavily labeled than upper layer III. Labeling in R in S101 is uniformly distributed in all cortical layers and probably is a result of diffusion of label from the injection site. Two different locations within the caudomedial field are heavily labeled. One area appears medial to the injection site (Fig. 4.12, S76) and the other is located more posteriorly (S41). All

---

Fɪɢ. 4.9. (continued)
with medial at the right and show labeling in the ipsilateral hemisphere. In E and F, medial is at the left and labeling in the contralateral hemisphere is shown. Bar in A indicates one mm in all six photomicrographs. A, three patches of label in RM resulting from an injection confined to A I (75-155M). B, labeling in AL, RM and R resulting from an injection confined to A I. Lateral arrow marks AL-R border. Medial arrow marks R–RM border (75-158M). C, labeling in A I following an injection confined to the rostral field. Arrow marks A I-PL border (75-153M). D, labeling in CM following an injection confined to A I. Arrow marks A I-CM border, with CM on the right and the injection site in A I on the left (75-155M). E, labeling in R resulting from an injection confined to A I in the contralateral hemisphere. Arrows mark medial and lateral boundaries of R (75-158M). F, labeling in A I resulting from an injection confined to A I of the contralateral hemisphere. Medial arrow marks the CM–A I border. Lateral arrow marks the A I–PL border (75-155M). From (27).

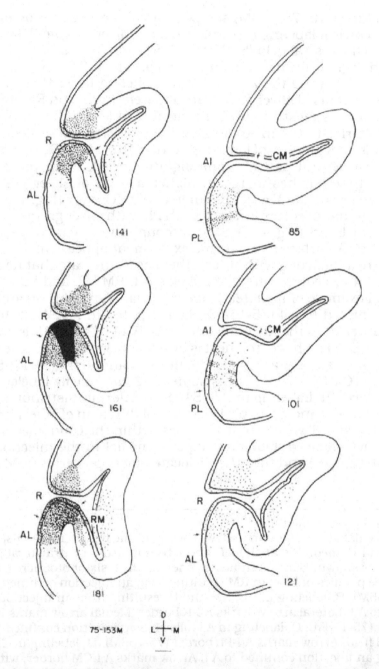

FIG. 4.10. Ipsilateral cortical projection from the rostral field. The distribution of silver grains is illustrated on line drawings of seven oblique sections through the superior temporal gyrus. Sections were cut in a plane perpendicular to the length of the Sylvian fissure. The lower left section is the most rostral section and the upper right is the most caudal.

cortical layers are labeled in both locations, but there is a slightly greater density of silver grains in layer IV and upper layer III than in other layers (Fig. 4.11D). The posterolateral field also receives a projection (Fig. 4.12, S31). Label is most dense in layer IV, somewhat less dense in layers I to III and relatively sparse in layers V and VI. A surface map for this experiment appears in Fig. 4.13. The results of four other experiments also suggest that neurons in A I project to fields R, CM, PL and RM. A projection from A I to AL is occasionally seen and labeling is such cases is quite light.

The results of experiments described in Figs. 4.10 and 4.12, along with results of an additional A I (75-11M) and R (75-153M) injection, are summarized in the form of surface maps in Fig. 4.13. Here the folded cortical surface of the superior temporal plane and gyrus has been represented as a flattened two-dimensional surface. Each map is oriented with anterior at the bottom and medial at the left. Solid heavy lines indicate cytoarchitectonic borders. The injection site is indicated in solid black. The area of diffuse labeling around the injection site is not shown. Stippling indicates areas of cortex in which transported label was found, although the density of stippling is not correlated with density of labeling. Arrows and numbers indicate the level of the corresponding sections in Figs. 4.10 and 4.12. As a result of an R injection in experiment 75-17M, two or more discrete patches of label are evident in each of fields AL, RM and A I. An additional laterally located patch lies near the A I-PL border. In a second injection located in the anterior portion of R(75-153M), AL and RM each contain a single patch of label. Not shown in this map is a small labeled area located approximately 1 mm rostral to the patch of label in RM. Whether this area is located in RM could not be determined, since the oblique cut through the cortex in this region obscured the cytoarchitecture. In anterolateral A I, three elongated parallel strips of label are seen, which merge posteriorly. Another patch of label appears at the A I-PL border.

Although in both experiments each of fields A I, PL, AL and RM contains labeled fiber terminals, the patterns of projections in the two experiments appear quite different. In the case of 75-17M, the greatest areal extent of the labeling is present in fields AL and RM,

---

FIG. 4.10. (continued)

Each is oriented with dorsal at the top and medial at the right. Bar indicates 1 mm. The labeled region surrounding the site of the injection of ^{3}H-proline is shown in solid black. This injection is confined to the rostral field. Arrows indicate borders between cytoarchitectonic fields. Density of dots indicates the relative density of autoradiographic label in these sections. Experiment 75-153M. From (27).

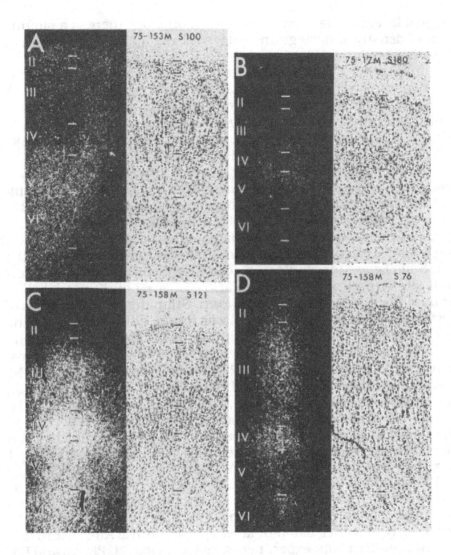

FIG. 4.11. Dark and bright field photomicrographs of the laminar distribution of silver grains in ipsilateral cortical fields resulting from injections in the rostral and primary fields. In each figure, the bright field photomicrograph of a section of cortex appears on the right adjacent to a dark field photomicrograph of the same sections on the left. All sections are oriented with the pial surface at the top. Magnification, 40 × . Sections are cut 25 μ thick in an oblique plane oriented perpendicular to the length of the Sylvian fissure and stained with cresylecht violet. Short lines indicate the boundaries of the six cortical layers designated by Roman numerals. A, labeling in A I resulting from an injection in R (75-153M). B, labeling in RM resulting from an injection in R (75-17M). C, labeling in R resulting from an injection confined to A I (75-158M). D, labeling in CM resulting from an injection confined to A I (75-158M). From (27).

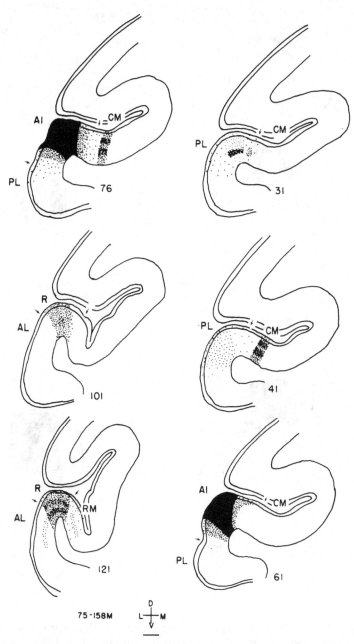

FIG. 4.12. Ipsilateral cortical projection resulting from an injection confined to the primary field (75-158M). See Fig. 4.10 for further details. From (27).

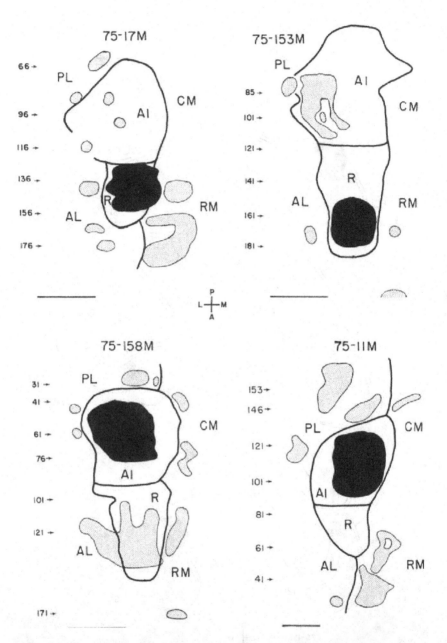

FIG. 4.13. Two-dimensional surface maps depicting the topography of projections resulting from single injections in R (75-17M, 75-153M) and A I (75-158M, 75-11M). In each case solid black lines represent the cytoarchitectonic boundaries of A I and R. The black area represents the injection site. Stippled areas represent regions of cortex receiving projections from the injection site. Each map is oriented with anterior at the bottom and medial at the right. Locations of sections illustrated in Figs. 4.10 and 4.12 are indicated by arrows. Bar represents 1 mm. From (27).

with relatively small areas labeled in fields A I and PL. On the other hand, the greatest areal extent of labeling in 75-153M is present in A I, with relatively small areas of PL, AL and RM containing label.

Following an A I injection (75-158M), two patches of label are evident in RM; the more posterior one is elongated in the antero-posterior dimension. A large labeled area is seen anteriorly in R, which extends into AL and breaks up into two elongated labeled areas more posteriorly in R. Four patches of label are evident in PL, two lateral to A I and two posterior to A I. CM contains two patches of label both elongated in the anteroposterior dimension. Following another A I injection (75-11M), one labeled area is seen in AL, two complexly shaped areas in RM, one in CM and three in PL. Labeling in R was probably masked by diffusion from the injection site in this case. In both cases, fields AL, RM, PL and CM receive projections from A I, although differences in the pattern of projections are seen in the two experiments. For example, in 75-158M four patches of label are evident in PL. In 75-11M, three labeled areas are seen in PL, but the labeled patches cover a larger area than in the previous case. One labeled patch is evident posteriorly in CM in 75-11M, whereas, in 75-158M, CM contains two labeled patches, one located posteriorly and one anteriorly. Although AL is labeled in both cases, only a small anterior patch is seen in 75-11M, whereas in 75-158M the labeled area in AL is much larger and apparently continuous with label in R. Two areas are labeled in RM in both cases, but the two patches are widely separated in 75-158M, and the total area of RM labeled is much greater in 75-11M.

8. Contralateral Cortical Projections

Widespread portions of the temporal lobe receive commissural input. Following section of all commissural pathways in monkey, Ebner and Myers (23) found temporal lobe degeneration only in the depths of superior temporal sulcus and superior parts of the superior and inferior temporal gyri. After sectioning the corpus callosum in rhesus monkey, Karol and Pandya (45) found moderate degeneration in the middle and posterior thirds of the superior temporal gyrus. Areas A I, proA, paAc (CM in owl monkey), paAr (R), paAlt (PL) and paI (RM) all contain degeneration after corpus callosum lesions (72).

In the owl monkey, more detailed information is available on the projection of the individual auditory fields. Figure 4.14 illustrates the pattern of labeling of the contralateral hemisphere, that resulted from an injection confined to A I (75-158M). The ipsilateral pattern of labeling resulting from this injection is seen in Figs. 4.12

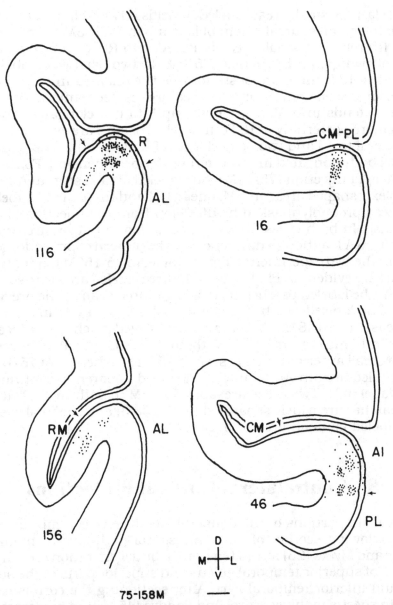

75-158M

FIG. 4.14. Distribution of silver grains in the auditory areas re-sulting from an injection confined to A I of the contralateral hemisphere (75-158M). Sections are oriented with medial at the left. See Fig. 4.10 for further details. From (27).

and 4.13. Silver grains are evident in the rostromedial field (S156), with moderate labeling in layer IV and a few scattered grains in lay-ers II and III. In the rostral field, silver grains are concentrated in upper layer III and layer IV (Fig. 4.14, S116; Fig. 4.9E). Moderate la-beling is present in layers V and VI and lower layer III, whereas layers I and II are lightly labeled. The primary field contains two patches or columns of label (S46) and the laminar distribution of silver grains in A I is similar to that in R. Most posteriorly (SI6), label is evident near the CM–PL border. Here silver grains are found in all layers, but are somewhat more concentrated in layers III and IV. This is the only case in which label was seen in CM or PL. Thus, these results indicate that A I projects contralaterally via the corpus callosum to layer IV of RM and to upper layer III and layer IV of R and A I.

The rostral field appears to project to fields A I, R and RM in the contralateral hemisphere. Figure 4.15 illustrates the result of an injection confined to the rostral field (77-54M). In R, a patch of label is evident (S61), in which silver grains are most heavily distributed in layer IV and upper layer III. The remaining layers are much more sparsely labeled. This patch disappears (S37) and more posteriorly a patch of label appears within R, near its border with A I (S25). Here again, label is heaviest in layer IV and upper layer III. Followed caudally, this patch breaks up into two distinct columns of label within A I (S13) with the same laminar distribution of silver grains. In section 1, heavy label is evident in layer IV and upper layer III of A I, with a few silver grains in other layers. In this experiment, light to moderate label was also found in layer IV of RM (S49). Thus, the contralateral projection of R appears very similar to that of A I, projecting primarily to layer IV and upper layer III of both R and A I. Layer IV of RM also receives some input from the contralateral rostral field.

In general, the commissural connections, that we have described in the owl monkey, terminate primarily in upper layer III and layer IV. Following callosal section in the rhesus monkey, fiber degeneration was heaviest in layers III and IV, lighter in layers V and VI and absent in I and II (38, 39), a finding that was confirmed in auditory cortex by Karol and Pandya (45). In a Fink-Heimer study, Pandya and Sanides (72) found that auditory cortex lesions in the rhesus monkey resulted in degeneration in the contralateral hemisphere, primarily in layers I, II, III and IV in A I and layer IV in other auditory fields. Thus, in both the owl monkey and rhesus monkey, commissural fibers appear to terminate primarily in layers III and IV of auditory cortex.

In the owl monkey, commissural projections from A I reach the opposite hemisphere via a different route than commissural projections from R. In experiments in which animals were allowed to survive for 5–7 days, labeled axons are evident in the corpus callosum following A I injections. No labeled axons are seen in the anterior commissure. On the other hand, following R injections, labeled axons are seen in the anterior commissure, but none in the corpus callosum. In primates, the rostral portions of the temporal lobes ap-

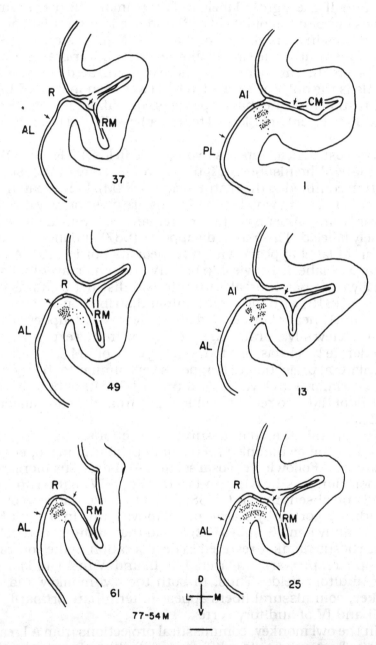

FIG. 4.15. Distributions of silver grains in the auditory areas resulting from an injection confined to R of the contralateral hemisphere (77-54M). Sections are oriented with medial at the right. See Fig. 4.10 for further details. From (27).

pear to be connected through the anterior commissure, whereas more caudal portions are connected through the corpus callosum. A Marchi study in monkey (30) indicated that, although the anterior limb of the anterior commissure connected olfactory structures, the posterior limb linked the middle temporal gyri. Ebner and Myers (23) stated that the anterior commissure appeared to link the inferior temporal convolutions in monkey. In the squirrel monkey, the rostral fourth of the superior temporal gyrus, the depth of the superior temporal sulcus and rostral lower bank of the Sylvian fissure, as well as the rostral third of the middle and inferior temporal gyri, are connected via the anterior commissure (71).

The commissural arrangement in the owl monkey contrasts with that in the rhesus monkey, in which neurons in association areas corresponding to RM, R, CM and PL project to the contralateral hemisphere in the corpus callosum (72). At present, it appears that the split auditory commissural projection seen in the owl monkey and the completely callosal connection of auditory areas seen in the rhesus monkey represent different plans of organization of auditory commissural systems.

The results of experiments described in Figs. 4.14 and 4.15, along with the results of an additional A I injection (75-155M), are summarized in the form of two-dimensional surface maps in Fig. 4.16. In experiment 75-158M, label is evident in the contralateral RM and AL following an A I injection. A large labeled area is present in the posterior part of contralateral R, with label extending into RM. A large labeled area is also evident in the posterior part of contralateral A I. Followed posteriorly, this patch breaks up into two columns of label elongated in the anteroposterior dimension. Two more small patches of label are seen anteriorly in A I. Labeled areas are also seen contralaterally in CM and at the CM-PL border. Following a second A I injection (75-155M), label is evident contralaterally in RM, R and CM. Label in contralateral A I again appears as several distinct columns of label elongated anteroposteriorly. An additional labeled area, located caudally, extends across the A I-CM border. Following an injection in the rostral field (77-54M), a labeled area is seen contralaterally in the anterior portion of R. This label extends into RM. Label is also present contralaterally in anterior A I and, in some sections, this labeled area breaks up into two discrete columns.

We have demonstrated that single injections in the primary and rostral fields in owl monkey result in labeling of two or more discrete columns of tissue in the ipsilateral and contralateral target fields. The patchy character of commissural corticocortical projec-

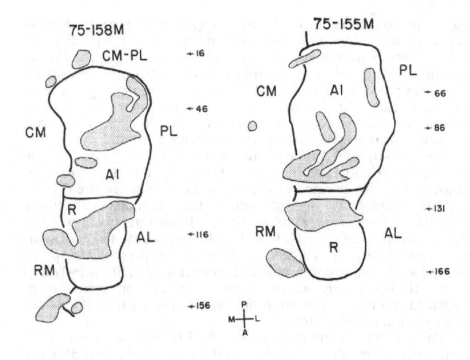

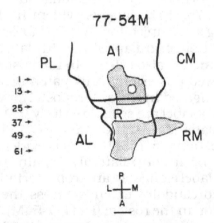

FIG. 4.16. Two-dimensional surface maps depicting the topography of projection sites in the contralateral hemisphere resulting from single injections in A I (75-158M, 75-155M) and R (77-54M). In each case, solid black lines represent the cytoarchitectonic boundaries of A I and R. Stippled areas represent regions of cortex receiving projections from the injection site in the contralateral hemisphere. Each map is oriented with anterior at the bottom. Medial is at the left in 75-158M and 75-155M and at the right in 77-54M. Bar represents 1 mm. Locations of sections illustrated in Figs. 4.14 and 4.15 are indicated by arrows. From (27).

tions is evident in other species of animals as well. After corpus callosum section in the rhesus monkey, Karol and Pandya (45) described three longitudinal bands of silver degeneration separated by clear areas in auditory cortex. Callosal section and A I lesions in rhesus monkey produced fiber degeneration primarily in the medial portion of the contralateral A I (70, 72). In studies of the commissural connections of the auditory areas in cat and raccoon, it was demonstrated that there are portions of A I that receive little callosal input (20, 24). In the high frequency representation of A I in the cat, callosal axon terminals aggregate into columns that correspond to the borders of physiologically defined binaural interaction columns (8, 35). It is possible that the columnar labeling seen here in the owl monkey reflects a similar physiological arrangement. A similar columnar or patchy pattern of ipsilateral and contralateral labeling in target fields occurs following injections in auditory (36), somatic motor (41–43, 47), visual (58, 94) and other cortical areas (31) in monkeys and carnivores.

In owl monkey, low-best frequencies are represented rostrolaterally and high-best frequencies caudomedially in both A I and R (37). The anterior portion of R projects to the ipsilateral anterolateral portion of A I in 75-153M (Fig. 4.13) and to the anterolateral portions of A I and R contralaterally in 77-54M (Fig. 4.16). On the other hand, the posterior part of A I projects to the posterior portions of A I and R contralaterally in 75-158M (Fig. 4.16). These results suggest that similar portions of the frequency representation in R and A I in both hemispheres are interconnected. These findings are consistent with results of combined anatomical and electrophysiological mapping experiments in the cat (36).

Figure 4.17 presents a summary of the projections of fields R and A I and of the laminar distribution of silver grains in the granular and supragranular layers of the target fields. The pattern seen in subgranular layers was not included, because this was more variable and often was continuous with labeling of the white matter. Neurons in R project to fields RM and A I in both hemispheres and to fields AL and PL in the ipsilateral hemisphere. In addition, the rostral fields in both hemispheres are connected. Neurons in the primary field project to AL, RM, R, CM and PL ipsilaterally and to RM, R and A I in the contralateral hemisphere. Corticocortical axon terminations are concentrated in layer IV of fields RM and AL and upper layer III and layer IV of R and CM. In the primary field, axon terminals of neurons, whose cell bodies lie in A I in the opposite hemisphere, are concentrated in upper layer III and layer IV; axon terminals of neurons located in field R of the same hemisphere are concentrated in layers I and II. Layer IV of PL contains the greatest

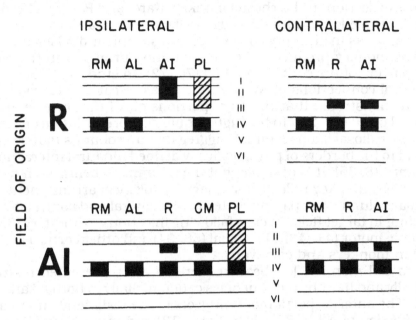

FIG. 4.17. Summary diagram of the cortical projections from A I and R. The projections from fields R and A I are shown in the upper and lower rows respectively. Projections to ipsilateral fields are depicted in the column on the left and projections to contralateral fields are shown in the column on the right. Densely labeled cortical laminae are shown as solid black; moderately labeled laminae contain diagonal lines. Only labeling of granular and supragranular layers is illustrated. From (27).

concentration of corticocortical axon terminals; the supragranular layers contain a somewhat lower concentration.

Based on connectivity patterns, it is possible to describe certain correspondences between A I and R in the owl monkey and the auditory areas of the cat. Recent connectivity studies, based on transport of tritiated amino acids and/or horseradish peroxidase, have indicated that cat A I is reciprocally connected with the ventral division of the medial geniculate (2, 54, 65, 93). Cat A I also appears to be connected with the magnocellular nucleus. The ventral and magnocellular nuclei in cat correspond cytoarchitectonically to the principal and magnocellular nuclei, respectively, in owl monkey (26). On the basis of these connectional similarities, A I in owl monkey corresponds to A I in cat. Examination of corticocortical connectivities in cat and monkey is unlikely to aid in establishing correspondences, since most auditory cortical fields appear to be reciprocally connected (20, 21, 36).

Which field (or fields) in cat might correspond to R in owl monkey? Recent microelectrode mapping studies of Knight (46) and Reale and Imig (76) have indicated that the physiological organization of auditory cortex outside A I in cat is somewhat different from that originally described by Woolsey (95, 96). At least three tonotopically organized fields are present outside A I. The anterior auditory field (A) includes portions of the suprasylvian fringe, whereas the posterior (P) and ventroposterior (VP) fields correspond to portions of EP. Since R displays a tonotopic organization, we would expect it to correspond to a tonotopically organized field in cat (i.e., A, P or VP). The anterior field in cat is connected with the deep dorsal division of the medial geniculate, but not with the ventral (principal) division (2, 65). On the other hand, field P is connected with both the dorsal and principal divisions of the medial geniculate (28, 65), as is R. The connections of VP have not been worked out. On the basis of the limited evidence available, it appears that R in the owl monkey may correspond to P in the cat. By comparing tonotopic maps of the cat and owl monkey, Woolsey (97) reached a similar conclusion.

We can now summarize the overall output patterns of A I and R. A I projects subcortically to the principal and magnocellular nuclei of the medial geniculate and to the laminated and dorsomedial portions of the central nucleus of the inferior colliculus.

The central nucleus of the inferior colliculus, principal nucleus of the medial geniculate body and primary auditory cortex, A I, each display a precise tonotopic organization (25, 33, 37), which at the collicular and geniculate levels is related to a laminar arrangement of the dendrites and fibers (25, 61, 62). The central nucleus of the inferior colliculus projects to the principal nucleus of the medial geniculate (14, 15, 59, 60, 66, 68), which sends axons to A I (2, 12, 15, 32, 54, 56, 65, 67, 74, 78, 83, 93). A I, in turn, projects back directly upon both the principal division of the medial geniculate and the central nucleus of the inferior colliculus. Primary auditory cortex is thus in a position to influence its own input directly at the level of both colliculus and geniculate.

Both A I and R send sparse projections to MGM, and the magnocellular nucleus projects to these fields (12). In cat, monkey and tree shrew, the magnocellular nucleus receives a projection from the inferior colliculus (14, 22, 59, 60, 66, 68). Unlike the principal nucleus, the magnocellular nucleus also receives input from cervical spinal cord (6, 13), the spinal lemniscus (64) and lateral tegmental fiber systems (63) in cat and from the tegmentum and deep superior colliculus in tree shrew (68, 69). Fibers of the brachium of the inferior colliculus project to the lateral tegmentum and deep superior colliculus in monkey (60). MGM appears to be part of a multimodality system and is in a position to integrate information from a variety of ascending and descending sources.

The rostral field projects to the principal and magnocellular nuclei of the medial geniculate, as does A I, but sends an additional projection to the posterodorsal nucleus of the geniculate as well. Burton and Jones (12) have demonstrated that the posterodorsal nucleus projects to R.

The dorsal nucleus of the medial geniculate is not a target of the inferior colliculus in cat and monkey (59, 60, 66), although in tree shrew the deep dorsal division receives fibers from the roof of the inferior colliculus (68) and the pericentral nucleus sends a projection to the caudal medial geniculate (15). Instead, the dorsal nucleus gets its input from the lateral tegmental fiber system (22, 63) and from the tegmentum and deep superior colliculus in tree shrew (68, 69).

R projects to the dorsomedial portion of the central nucleus of the inferior colliculus, as well as the pericentral and external nuclei. These cell groups may project to the dorsal nucleus of the medial geniculate via the tegmental system and deep superior colliculus.

A I projects to fields R, PL, RM and CM in the ipsilateral hemisphere and to fields A I, R and RM in the contralateral hemisphere. R projects to fields A I, PL and RM ipsilaterally and to A I, R and RM contralaterally. Although A I projects strongly upon field CM, we have never seen labeling in CM after an R injection. Corticocortical axons originating in R appear to terminate primarily in layers I and II in ipsilateral A I, whereas axons of A I neurons are distributed to layers III and IV in ipsilateral R.

The differences in connectivities that exist between A I and R imply that these two fields are processing different overall patterns of information. Very few studies have examined the discharge characteristics of single cells in the various regions of primate auditory cortex (7, 9). Information of this type is essential, if we are to analyze the functional differences between these fields.

Abbreviations

A   Anterior
AL   Anterolateral cortical field
ALM   Anterolateral margin
A I   Primary auditory field

BCI   Brachium of the inferior colliculus
BCS   Brachium of the superior colliculus
CM   Caudomedial cortical field
DM   Dorsomedial region of the inferior colliculus
DNLL   Dorsal nucleus of the lateral lemniscus
IC   Inferior colliculus
ICC   Central nucleus of the inferior colliculus
ICP   Pericentral nucleus of the inferior colliculus
ICX   External nucleus of the inferior colliculus
IP   Inferior pulvinar
L   Lateral
LGN   Lateral geniculate nucleus
LIM   Nucleus limitans
LL   Lateral lemniscus
M   Medial
MGB   Medial geniculate body
MGD   Dorsal nucleus of the medial geniculate
MGDA   Anterior division of the dorsal nucleus of the medial geniculate
MGDP   Posterior division of the dorsal nucleus of the medial geniculate
MGM   Magnocellular nucleus of medial geniculate
MGP   Principal nucleus of medial geniculate
P   Posterior
PAG   Periaqueductal gray
PL   Posterolateral cortical field
PMZ   Posterior marginal zone
Po   Posterior group
Pu   Pulvinar
PuM   Medial pulvinar
R   Rostral cortical field
SC   Superior colliculus
SG   Suprageniculate nucleus

References

1. AKERT, K., WOOLSEY, C. N., DIAMOND, I. T., AND NEFF, W. T. The cortical projection area of the posterior pole of the medial geniculate body in *Macaca mulatta. Anat. Rec.*, 133: 242, 1959.
2. ANDERSEN, R. A. *Patterns of Connectivity of the Auditory Forebrain of the Cat.* PhD dissertation. The University of California at San Francisco, 1979.

3. BAILEY, P., GAROL, H. W., AND MCCULLOCH, W. S. Cortical origin and distribution of corpus callosum and anterior commissure in chimpanzee (*Pan satyrus*). *J. Neurophysiol.*, 4: 564–571, 1941.

4. BAILEY, VON BONIN, G., GAROL, H. W., AND MCCULLOCH, W. S. Functional organization of temporal lobe of monkey (*Macaca mulatta*) and chimpanzee (*Pan satyrus*). *J. Neurophysiol.*, 6: 121–128, 1943.

5. BERMAN, A. L., AND JONES, E. G. *The Thalamus and Basal Telencephalon of the Cat. A Cytoarchitectonic Atlas with Sterotaxic Coordinates.* Madison: University of Wisconsin Press. In press, 1981.

6. BOIVIE, J. The termination of the cervicothalamic tract in the cat. An experimental study with silver impregnation methods. *Brain Res.*, 19: 333–360, 1970.

7. BRUGGE, J. F. Mechanisms of coding information in the auditory system. *Acta Symbolica*, 6 (2): 35–63, 1975.

8. BRUGGE, J. F., AND IMIG, T. J. Some relationships of binaural response patterns of single neurons to cortical columns and interhemispheric connections of auditory area A I of cat cerebral cortex. In: *Evoked Electrical Activity in the Auditory Nervous System*, edited by R. F. NAUNTON AND C. FERNANDEZ, New York: Academic Press, 1978, pp. 487–503.

9. BRUGGE, J. F., AND MERZENICH, M. M. Responses of neurons in auditory cortex of the macaque monkey to monaural and binaural stimulation. *J. Neurophysiol.*, 36: 1138–1158, 1973.

10. BUCY, P. C., AND KLÜVER, H. Anatomic changes secondary to temporal lobectomy. *Arch. Neurol. Psychiat.*, 44: 1142–1146, 1940.

11. BUCY, P. C., AND KLÜVER, H. An anatomical investigation of the temporal lobe in the monkey (*Macaca mulatta*). *J. Comp. Neurol.*, 103: 151–251, 1955.

12. BURTON, H., AND JONES, E. G. The posterior thalamic region and its cortical projection in new world and old world monkeys. *J. Comp. Neurol.*, 168: 249–302, 1976.

13. CARSTENS, E., AND TREVINO, D. L. A projection from upper cervical spinal cord to ipsilateral thalamus in cat and monkey: A new somatosensory relay from the body? *Neuroscience Abstr.*, 2: 971, 1976.

14. CASSEDAY, J. H., AND DIAMOND, I. T. Connections from the inferior colliculus to auditory cortex via the medial geniculate in *Tupaia glis*. *Anat. Rec.*, 178: 324, 1974.

15. CASSEDAY, J. H., DIAMOND, I. T., AND HARTING, J. K. Auditory pathways to the cortex in *Tupaia glis*. *J. Comp. Neurol.*, 166: 303–340, 1976.

16. CHOW, K. L. A retrograde cell degeneration study of the cortical projection field of the pulvinar in the monkey. *J. Comp. Neurol.*, 93: 313–340, 1950.

17. CURTIS, H. J. Intercortical connections of corpus callosum as indicated by evoked potentials. *J. Neurophysiol.*, 3: 407–413, 1940.

18. CURTIS, H. J., AND BARD, P. Intercortical connections of corpus callosum as indicated by evoked potentials. *Amer. J. Physiol.*, 126: P473, 1939.

19. DE VITO, J. L., AND SIMMONS, D. M. Some connections of the posterior thalamus in monkey. *Exptl. Neurol.*, 51: 347–362, 1976.

20. DIAMOND, I. T., JONES, E. G., AND POWELL, T. P. S. Interhemispheric fiber connections of the auditory cortex of the cat. *Brain Res.*, 11: 177–193, 1968a.

21. DIAMOND, I. T., JONES, E. G., AND POWELL, T. P. S. The association connections of the auditory cortex of the cat. *Brain Res.*, 11: 560–579, 1968b.

22. DIAMOND, I. T., JONES, E. G., AND POWELL, T. P. S. The projection of the auditory cortex upon the diencephalon and brain stem in the cat. *Brain Res.*, 15: 305–340, 1969.

23. EBNER, F. F., AND MYERS, R. E. Commissural connections in the neocortex of monkey. *Anat. Rec.*, 142: 229, 1962.

24. EBNER, F. F., AND MYERS, R. E. Distribution of corpus callosum and anterior commissure in cat and raccoon. *J. Comp. Neurol.*, 124: 353–366, 1965.

25. FITZPATRICK, K. A. Cellular architecture and topographic organization of the inferior colliculus of the squirrel monkey. *J. Comp. Neurol.*, 164: 185–208, 1975.

26. FITZPATRICK, K. A., AND IMIG, T. J. Projections of auditory cortex upon the thalamus and midbrain in the owl monkey. *J. Comp. Neurol.*, 177: 537–556, 1978.

27. FITZPATRICK, K. A., AND IMIG, T. J. Auditory cortico-cortical connections in the owl monkey. *J. Comp. Neurol.*, 192: 589–610, 1980.

28. FITZPATRICK, K. A., IMIG, T. J., AND REALE, R. A. Thalamic projections to the posterior auditory cortical field in cat. *Neuroscience Abstr.*, 3: 6, 1977.

29. FORBES, B. F., AND MOSKOWITZ, N. Projections of auditory responsive cortex in the squirrel monkey. *Brain Res.*, 67: 239–254, 1974.

30. FOX, C. A., FISHER, R. R., AND DISALVA, S. J. The distribution of the anterior commissure in the monkey (*Macaca mulatta*). *J. Comp. Neurol.*, 89: 245–277, 1948.

31. GOLDMAN, P. S., AND NAUTA, W. J. H. Columnar distribution of cortico-cortical fibers in the frontal association, limbic and motor cortex of the developing rhesus monkey. *Brain Res.*, 122: 393–413, 1977.

32. GRAYBIEL, A. The thalamocortical projection of the so-called posterior nuclear group: A study with anterograde degeneration methods in the cat. *Brain Res.*, 49: 229–244, 1973.

33. GROSS, N. B., LIFSCHITZ, W. S., AND ANDERSON, D. J. The tonotopic organization of the auditory thalamus of the squirrel monkey (*Saimiri sciureus*). *Brain Res.*, 65: 323–332, 1974.

34. HURST, E. M. Some cortical association systems related to auditory functions. *J. Comp. Neurol.*, 112: 103–119, 1959.

35. IMIG, T. J., AND BRUGGE, J. F. Sources and terminations of callosal axons related to binaural and frequency maps in primary auditory cortex of the cat. *J. Comp. Neurol.*, 182: 637–660, 1978.

36. IMIG, T. J., AND REALE, R. A. Patterns of the cortico-cortical connections related to tonotopic maps in cat auditory cortex. *J. Comp. Neurol.*, 192: 1980.

37. IMIG, T. J., RUGGERO, M. A., KITZES, L. M., JAVEL, E., AND BRUGGE, J. F. Organization of auditory cortex in the owl monkey (*Aotus trivirgatus*). *J. Comp. Neurol.*, 171: 111–128, 1977.

38. JACOBSON, S., AND MARCUS, E. M. Comparison of the laminar distribution of callosal synapses in the rat and rhesus monkey. *Anat. Rec.*, 163: 203, 1969.

39. JACOBSON, S., AND MARCUS, E. M. The laminar distribution of fibers of the corpus callosum; a comparative study in the rat, cat, rhesus monkey and chimpanzee. *Brain Res.*, 24: 517–520, 1970.

40. JASPER, H., AJMONE-MAR SAN, C., AND STOLL, J. Corticofugal projections to the brain stem. *Arch. Neurol. Psychiat.*, 67: 155–171, 1952.

41. JONES, E. G., BURTON, H., AND PORTER, R. Commissural and corticortical "columns" in the somatic sensory cortex of primates. *Science*, 190: 572–574, 1975.

42. JONES, E. G., COULTER, J. D., AND HENDRY, S. H. C. Intracortical connectivity of architectonic fields in the somatic sensory, motor and parietal cortex of monkeys. *J. Comp. Neurol.*, 181: 291–348, 1978.

43. JONES, E. G., COULTER, J. D., AND WISE, S. P. Commissural columns in the sensory-motor cortex of monkeys. *J. Comp. Neurol.*, 188: 113–136, 1979.

44. JONES, E. G., AND POWELL, T. P. S. Connexions of the somatic sensory cortex of the rhesus monkey. III. Thalamic connexions. *Brain*, 93: 37–56, 1970.

45. KAROL, E. A., AND PANDYA, D. N. The distribution of the corpus callosum in the rhesus monkey. *Brain*, 94: 471–486, 1971.

46. KNIGHT, P. L. Representation of the cochlea within the anterior auditory field (AAF) of the cat. *Brain Res.*, 130: 447–467, 1977.

47. KÜNZLE, H. Cortico-cortical efferents of primary motor and somatosensory regions of the cerebral cortex in *Macaca fascicularis*. *Neuroscience*, 3: 25–39, 1978.

48. KUYPERS, H. G. J. M., AND LAWRENCE, D. G. Cortical projections to the red nucleus and the brain stem in the rhesus monkey. *Brain Res.*, 4: 151–188, 1967.

49. LE GROS CLARK, W. E. The thalamic connections of the temporal lobe of the brain in the monkey. *J. Anat., London*, 70: 447–464, 1936.

50. LE GROS CLARK, W. E., AND NORTHFIELD, D. W. C. The cortical projection of the pulvinar in the macaque monkey. *Brain*, 60: 126–142, 1937.

51. LOCKE, S. The projection of the medial pulvinar of the macaque. *J. Comp. Neurol.*, 115: 155–169, 1960.

52. MEHLER, W. R., FEFERMAN, M. E., AND NAUTA, W. J. H. Ascending axon degeneration following anterolateral cordotomy. An experimental study in the monkey. *Brain*, 83: 718–750, 1960.

53. MCCULLOCH, W. S., AND GAROL, H. W. Cortical origin and distribution of the corpus callosum and anterior commissure in the monkey (*Macaca mulatta*). *J. Neurophysiol.*, 4: 555–563, 1941.

54. MERZENICH, M. M., AND COLWELL, S. A. Spatially ordered convergent projection from the auditory thalamus to and from A I in the cat. *J. Acoust. Soc. Amer.*, 57: S 55, 1975.

55. MERZENICH, M. M., AND BRUGGE, J. F. Representation of the cochlear partition on the superior temporal plane of the macaque monkey. *Brain Res.*, 50: 275–296, 1973.

56. MESULAM, M.-M., AND PANDYA, D. P. The projections of the medial geniculate complex within the Sylvian fissure of the rhesus monkey. *Brain Res.*, 60: 315–333, 1973.

57. METTLER, F. A. Corticofugal fiber connections of the cortex of *Macaca mulatta*. The temporal region. *J. Comp. Neurol.*, 63: 25–47, 1935.

58. MOORE, R. Y., AND GOLDBERG, J. M. Ascending projections of the inferior colliculus in the cat. *J. Comp. Neurol.*, 121: 109–135, 1963.

59. MOORE, R. Y., AND GOLDBERG, J. M. Projections of the inferior colliculus in the monkey. *Exptl. Neurol.*, 14: 429–438, 1966.

60. MONTERO, V. M. Patterns of connections from the striate cortex to cortical visual areas in superior temporal sulcus of macaque and middle temporal gyrus of owl monkey. *J. Comp. Neurol.*, 189: 45–59, 1980.

61. MOREST, D. K. The neuronal architecture of the medial geniculate body of the cat. *J. Anat., London*, 98: 611–630, 1964.

62. MOREST, D. K. The laminar structure of the medial geniculate body of the cat. *J. Anat., London*, 99, 1: 143–160, 1965a.

63. MOREST, D. K. The lateral tegmental system of the midbrain and the medial geniculate body: study with Golgi and Nauta methods in cat. *J. Anat., London*, 99: 611–634, 1965b.

64. NAUTA, W. J. H., AND KUYPERS, H. G. J. M. Some ascending pathways in the brain stem reticular formation. In: *Reticular Formation of the Brain*, edited by H. H. JASPER AND L. D. PROCTOR. Henry Ford Hospital Symposium. Boston: Little, Brown, 1958, pp. 3–30.

65. NIIMI, K., AND MATSUOKA, H. *Thalamocortical Organization of the Auditory System in the Cat Studied by Retrograde Axonal Transport of Horseradish Peroxidase. Adv. Anat. Embryol. Cell Bio.* 57: 1–58, 1979.

66. VAN NOORT, J. *The Structure and Connections of the Inferior Colliculus.* Leiden: Van Gorcum, 1969.

67. OLIVER, D. L. Cytoarchitecture and connections of the thalamocortical auditory system in the tree shrew. *Tupaia glis. Anat. Rec.*, 178: 430, 1974.

68. OLIVER, D. L. Midbrain projections to the medial geniculate body and their relationship to corticofugal projections of the auditory cortex. *Neuroscience Abstracts*, 2: 8, 1976.

69. OLIVER, D. L., AND HALL, W. C. Subdivisions of the medial geniculate body in the tree shrew (*Tupaia glis*). *Brain Res.*, 86: 217–227, 1975.

70. PANDYA, D. N., HALLETT, M., AND MUKHERJEE, S. K. Intra- and interhemispheric connections of the neocortical auditory system in the rhesus monkey. *Brain Res.*, 14: 49–65, 1969.

71. PANDYA, D. N., KAROL, E. A., AND LELE, P. P. The distribution of the anterior commissure in the squirrel monkey. *Brain Res.*, 49: 177–180, 1973.

72. PANDYA, D. N., AND SANIDES, F. Architectonic parcellation of the temporal operculum in rhesus monkey and its projection pattern. *Z. Anat. Entwickl. Gesch.*, 139: 121–161, 1973.

73. PETR, R., HOLDEN, L. B., AND JIROUT, J. The efferent intercortical connections of the superficial cortex of the temporal lobe (*Macaca mulatta*). *J. Neuropath. Exptl. Neurol.*, 8: 100–103, 1949.

74. RACZKOWSKI, D., DIAMOND, I. T., AND WINER, J. Organization of the thalamocortical auditory system in the cat studied with horseradish peroxidase. *Brain Res.*, 101: 345–354, 1975.

75. RAMÓN Y CAJAL, S. *Histologie du Système Nerveux de l'Homme et des Vertébrés*, Vol. II. Consejo Superior de Investigaciones Cientificas, Instituto Ramón y Cajal, Madrid, 1955 (Reprint of 1911 edition).

76. REALE, R. A., AND IMIG, T. J. Tonotopic organization in auditory cortex of the cat. *J. Comp. Neurol.*, 192: 265–292, 1980.

77. ROCKEL, A. J., AND JONES, E. G. The neuronal organization of the inferior colliculus of the adult cat. I. The central nucleus. *J. Comp. Neurol.*, 147: 11–60, 1973.

78. ROSE, J. E., AND WOOLSEY, C. N. Cortical connections and functional organization of the thalamic auditory system in the cat. In: *Biological and Biochemical Bases of Behavior*, edited by H. F. HARLOW AND C. N. WOOLSEY. Madison: University of Wisconsin Press, 1965.

79. RUNDLES, R. W., AND PAPEZ, J. W. Fibers and cellular degeneration following temporal lobectomy in the monkey. *J. Comp. Neurol.*, 68: 267–296, 1938.

80. SIMPSON, D. A. The projection of the pulvinar to the temporal lobe. *J. Anat., London*, 86: 20–28, 1952.

81. SIQUEIRA, E. B. The tempero-pulvinar connections in the rhesus monkey. *Arch. Neurol.*, 13: 321–330, 1965.

82. SIQUEIRA, E. B. The cortical connections of the nucleus pulvinaris of the dorsal thalamus in the rhesus monkey. *J. Hirnforsch.*, 10: 487–498, 1968.

83. SOUSA-PINTO, A. Cortical projections of the medial geniculate body in the cat. *Adv. Anat. Embryol. Cell Biol.*, 48: 2: 1–40, 1973.

84. SUGAR, O., FRENCH, J. D., AND CHUSID, J. G. Corticocortical connections of the superior surface of the temporal operculum in the monkey (*Macaca mulatta*). *J. Neurophysiol.*, 11: 175–184, 1948.

85. THOMPSON, W. H. Degenerations resulting from lesions of the cortex of the temporal lobe. *J. Anat. Physiol.*, 35: 147–165, 1901.

86. TROJANOWSKI, J. Q., AND JACOBSON, S. A combined horseradish peroxidase autoradiographic investigation of reciprocal connections between superior temporal gyrus and pulvinar in squirrel monkey. *Brain Res.*, 85: 347–353, 1975.

87. WALKER, A. E. An experimental study of the thalamocortical projection of the macaque monkey. *J. Comp. Neurol.*, 64: 1–39, 1936.

88. WALKER, A. E. The projection of the medial geniculate body to the cerebral cortex in the macaque monkey. *J. Anat., London*, 71: 319–331, 1937.

89. WALKER, A. E. *The Primate Thalamus*. Chicago: The University of Chicago Press, 1938.

90. WALKER, A. E., AND FULTON, J. F. The thalamus of the chimpanzee. III. Metathalamus, normal structure and cortical connections. *Brain*, 61: 250–268, 1938.

91. WALZL, E. M. Representation of the cochlea in the cerebral cortex. *Laryngoscope*, 57: 778–787, 1947.

92. WHITLOCK, D. G., AND NAUTA, W. J. H. Subcortical projections from the temporal neocortex in *Macaca mulatta*. *J. Comp. Neurol.*, 106: 183–212, 1956.

93. WINER, J. A., DIAMOND, I. T., AND RACZKOWSKI, D. Subdivision of the auditory cortex of the cat; the retrograde transport of horseradish peroxidase to the medial geniculate body and posterior thalamic nuclei. *J. Comp. Neurol.*, 176: 387–418, 1977.

94. WONG-RILEY, M. Columnar cortico-cortical interconnections within the visual system of the squirrel and macaque monkeys. *Brain Res.*, 162: 201–217, 1979.

95. WOOLSEY, C. N. Organization of the cortical auditory system; a review and a synthesis. In: *Neural Mechanisms of the Auditory and Vestibular Systems*, edited by G. F. RASMUSSEN AND W. F. WINDLE. Springfield, IL.: C. C Thomas, 1960, pp. 165–180.

96. WOOLSEY, C. N. Organization of the cortical auditory system. In: *Sensory Communication*, edited by W. A. ROSENBLITH. Cambridge, MA: The MIT Press, 1961, pp. 235–257.

97. WOOLSEY, C. N. Tonotopic organization of the auditory cortex. In: *Physiology of the Auditory System*, edited by M. B. SACHS. Baltimore: National Educational Consultants. 1971, pp. 271–282.

Chapter 5

Polysensory "Association" Areas of the Cerebral Cortex

Organization of Acoustic Input in the Cat

D. R. F. Irvine and D. P. Phillips

Neuropsychology Laboratory, Department of Psychology, Monash University, Clayton, Victoria, Australia

1. Introduction

1.1. Cortical Association Areas and Hypothesized Sources of Sensory Input

The multiple representations of the cochlea on the cat's ectosylvian cortex have been described by other contributors to this volume (see Imig et al., Merzenich, Colwell and Andersen, chapters 1 and 2). As Woolsey's (174) classical diagram of the cortical auditory areas (Fig.5.1A) indicates, however, acoustic input is not restricted

111

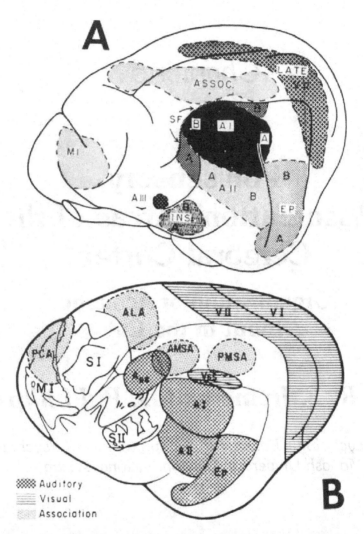

FIG. 5.1. A, Woolsey's (174) diagram summarizing his proposal concerning organization of cortical auditory system in cat. Five divisions containing complete representations of the cochlea were identified. Areas in "association" cortex (ASSOC) and precentral motor cortex (MI) were described as giving responses to click with 15-ms latency under chloralose anesthesia. (Reproduced courtesy of Charles C Thomas, Publisher, Springfield, Illinois). B, approximate locations of association response foci as determined by Thompson and colleagues and described in text. From Thompson et al. (159).

to these specific projection fields. In unanesthetized (18, 35, 38, 51, 56, 57, 121, 122, 139, 171) and chloralose-anesthetized (2, 21, 33, 36, 38, 49, 51, 52, 87, 139, 141, 147, 159, 160, 162, 171) cats, responses to auditory stimulation are also seen in a number of other discrete areas of the cerebral cortex. Detailed mapping of the distribution of click-evoked responses (34, 36, 159, 160) has identified three major regions of this kind (designated as "association" areas in Fig. 5.1B). One of these areas is situated on the medial suprasylvian gyrus (MSA), and is sometimes described as consisting of two discrete (anterior and posterior) foci. A second area is situated on the anterior lateral gyrus (ALA), and a third on pericruciate "sensorimotor" cortex (PCA). It should be emphasized that these areas have been defined physiologically and do not correspond directly with known architectonic fields; MSA and ALA appear to span contiguous parts of (at least) areas 5 and 7, while PCA corresponds approximately with area 6 (79).

Because there is convergence of acoustic, visual and somatic input to neurons in these areas, they have frequently been referred to as "polysensory" or "nonspecific" cortex. Most commonly they are designated "association" areas and, despite its functional connotations, that term will be used here. These terminological issues, and much of the early evidence concerning the characteristics of these cortical fields, have been reviewed in detail by Buser and Bignall (34).

One of the fundamental questions about the association areas has concerned the organization of their sensory input and their relationship to the sensory lemniscal systems. At the risk of oversimplification, two broadly different views on this question can be distinguished. One has been that the input is derived from a nonspecific projection system, parallel to, but independent of, the higher levels of the primary sensory pathways. This hypothesis was initially proposed by Albe-Fessard to account for somatic input to association cortex (8, 11), and was subsequently extended by Thompson (159, 160, 162) to visual and acoustic input. It states that polysensory convergence takes place in a subcortical association system, consisting of the brain-stem reticular formation (RF) and medial/intralaminar (M/IL) thalamus, which projects in an equivalent fashion to the cortical association fields. In opposition to this view, others have argued that considerable convergence takes place in the cortex itself and that the convergent input is derived from higher (thalamic and cortical) levels of the primary sensory systems (49, 51, 52). There is similar uncertainty concerning the sources of input to frontal and parietal polysensory areas of squirrel monkey (24–26, 119).

In the following sections of this chapter, evidence bearing on the acoustic input to the association fields in the cat will be examined. After a brief review of earlier lesion studies, more recent data on the acoustic properties of neurons in association cortex and in its putative input sources will be presented, and the implications of these data for the organization of acoustic input to the association fields will be discussed.

1.2. Lesion Studies of Acoustic Input and the Question of Specific Receptive-Field Properties

Two main lines of indirect evidence have been adduced in support of the nonspecific projection system hypothesis. One involves the similarity of responses and response interactions in the association fields and in the postulated subcortical projection system (157, 159, 160, 162). The other derives from the fact that cortical potentials elicited by single-pulse electrical stimulation of either RF (33, 128) or M/IL thalamus (8, 11) have the same spatial distribution as responses to peripheral sensory stimulation.

Attempts to provide a more direct test of the hypothesis have involved determination of the effects of lesions of, or other forms of interference with, those areas considered to be likely input sources. With reference to acoustic input, it has been reported that association responses are not eliminated by removal of all auditory cortex (36, 160) or by interruption of the brachium of the inferior colliculus (IC; 2). These results support the notion that acoustic input is derived from pathways independent of the higher levels of the auditory lemniscal system. Robertson and Thompson's (141) report that association responses are eliminated by lesions of mesencephalic RF or posteromedial thalamus appears to locate this independent pathway in the postulated nonspecific system. This interpretation is supported by a recent report that cooling of medial thalamus reduces the magnitude of click-evoked responses in the cortical association fields (122).

In contrast to these positive results, others have reported that lesion of the RF (22, 36) or medial thalamus (95) has no permanent effect on auditory association responses. Buser et al. (36) suggested that acoustic input to the asssociation fields was in fact derived from parts of the posterior group of thalamic nuclei, while Liu and Shen (105) reported that click responses in motor cortex were dependent on the integrity of the medial geniculate body (MG).

The discrepancies between these results are undoubtedly attributable in part to limitations of the lesion technique itself. In the midbrain and thalamus, it is difficult to make lesions that both completely eliminate, and are entirely restricted to, either the nonspecific system or the primary sensory pathway.

Recent studies of visual input to the suprasylvian association field, however, suggest that the equivocal results of lesion studies might also be attributable to the fact that these studies have routinely employed stimuli of low specificity. Dow and Dubner (47) described one class of neuron in anterior MSA that was sensitive to movement and had large receptive fields extending as far as 20° into the ipsilateral visual hemifield. The response of such neurons to movement in a given hemifield was eliminated by ablation of the primary visual cortex (area 17) contralateral to that hemifield (48). Of particular interest in the present context is the observation by Dubner and Brown (50) that, although response specificity in anterior MSA was eliminated by lesions of visual cortex, responses to diffuse illumination depended only on the integrity of the lateral geniculate body.

The most obvious implication of these results is the inadequacy of the nonspecific projection system hypothesis with respect to visual input to MSA. The data also, however, establish a more general, methodological point. It is clear that the involvement of primary cortex would not have been recognized if only diffuse visual stimulation had been employed. Clarification of the organization of sensory input to the cortical association fields is, therefore, unlikely to be achieved unless adequate attention is given to the detailed receptive-field properties of single neurons. The aim of the research described in the following sections of this chapter has been to provide a detailed characterization of the acoustic input to nonspecific cortex and to its putative input sources, as a first step in elucidating the organization of this input.

2. Methods

All of the experiments to be described employed adult cats, in most cases anesthetized with α-chloralose. Details of the preparation and of stimulating, recording and analytic procedures have been reported in detail elsewhere (86, 87, 126) and will be only briefly summarized here. The pinnae were removed and round-window electrodes were implanted bilaterally to monitor peripheral sensitivity. Tone- and noise-burst stimuli (10 ms rise-fall time; 250 ms duration) were presented using sealed systems incorporating probe-microphone assemblies. Gross visual and somatic stimuli (diffuse light flash and forepaw shock, respectively) were used for routine testing of polysensory convergence. Low rates of stimulation (1/2.5 s to 1/4 s) were used, because of the well-known rate sensitivity and lability of association cortex neurons. The activity of single neurons was recorded by conventional extracellular techniques and analysis of discharge characteristics was carried out by means of a Data General Nova computer. All response histograms and spike-count data are based on cumulated responses to 20 stimulus presentations.

3. Acoustic Properties of Neurons in Cortical Association Fields

3.1. General Response Properties

Observations were made on a total of 652 neurons in the three association fields (87). As indicated in Fig. 5.2A, approximately 70% of the neurons isolated in each of the areas were driven by acoustic stimulation. The majority of responsive units were also driven by visual and/or somatic stimulation, only a small proportion being unimodal (Fig.5.2B–D). The polysensory input to these neurons and other general characteristics, such as rate sensitivity and response lability, have been described in detail by others (34, 49, 51, 139, 171). The present account will, therefore, be restricted to their acoustic properties, and will summarize observations presented in greater detail elsewhere (87).

In chloralose-anesthetized cats, all acoustically activated units respond to tone- or noise-burst stimulation with simple onset responses. In those neurons that exhibit spontaneous activity, the onset response is followed by a long-lasting (200–300 ms) suppression of background activity. These characteristics of association cortex neurons are identical to those of onset cells in M/IL thalamus, and are illustrated by the thalamic response histograms in Fig.5.7A,B.

Minimum latencies to the first spike for neurons in the three association areas varied from 16 to 54 ms for those units on which precise measures could be obtained from response histograms. These figures are biased toward lower values because less-securely driven cells, for which response histograms were not routinely obtained, exhibited longer and more variable latencies. The latency data, and the restrictions placed on their interpretation by this sampling bias, are discussed in more detail in a later section (see Fig.5.9 and related text).

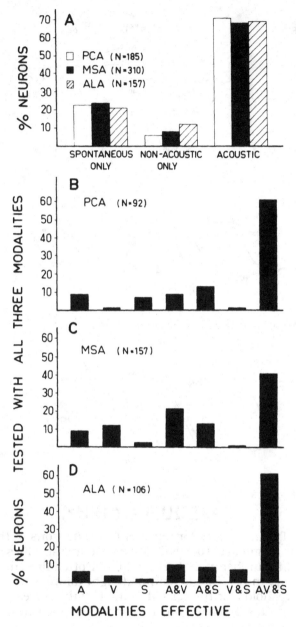

FIG. 5.2. A, proportions of neurons in each cortical association area that were driven acoustically (acoustic), driven only by nonacoustic stimulation (nonacoustic only), or that were not driven by any of the available stimuli (spontaneous only). B–D, proportions of those neurons tested with all three modalities that responded to auditory (A), visual (V), and somatosensory (S) stimulation. Categories are exclusive and conjunctions indicate responses to each of the modalities presented separately. In each case, N specifies the number of neurons on which proportions are based. From Irvine and Huebner (87).

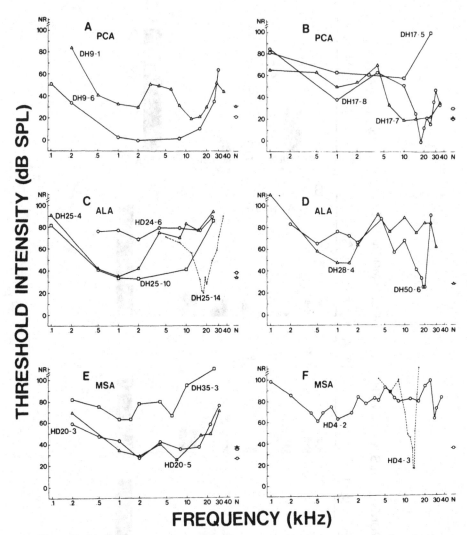

FIG. 5.3. Tuning curves for representative neurons in the three association areas. Numeral associated with each curve specifies cat and (following dash) unit number in that preparation. Dotted-line curves in C and F are for neurons in A I. Single points at right above N on abscissa are white-noise thresholds for the cells identified by the corresponding symbols. NR on ordinate indicates no response. Modified from Irvine and Huebner (87).

3.2. Frequency Tuning and Thresholds

Representative frequency-tuning curves for association cortex neurons are presented in Fig. 5.3. As these curves illustrate, the majority of neurons were very broadly tuned, responding over 5 to 6 octaves or more. That this broad tuning is not attributable to a general effect of chloralose anesthesia is indicated by two observations. First, control recordings from A I neurons in the same animals (e.g., units 25-14 and 4-3 in Fig.5.3C and F, respectively) showed the sharp tuning characteristic of most primary auditory cortical neurons in barbiturate and unanesthetized cats (1, 60, 68, 82, 127). Second, a small proportion of association cortex neurons also exhibited sharp tuning comparable to that of A I (units 17-8 and 50-6 in Fig.5.3B and D, respectively). Conventional methods of measuring tuning curve sharpness (Q_{10}; 100) are not applicable to broad irregular curves of the type obtained from most association cortex neurons. As an alternative, we have employed the response range (in octaves) at a level 40 dB above the most sensitive point on the curve. Breadth-of-tuning distributions obtained by the application of this measure are presented in Fig.5.4 and emphasize the broad tuning of the majority of association cortex neurons.

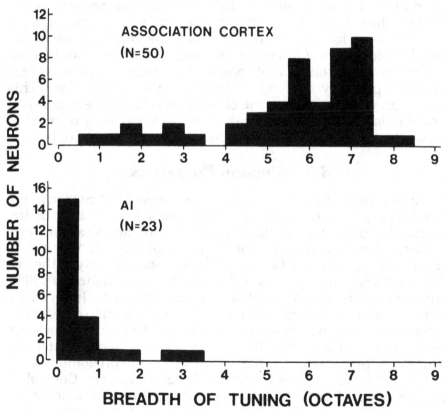

FIG. 5.4. Breadth-of-tuning distributions for neurons in association cortex and in A I. Index of breadth of tuning defined in text. A I distribution is based on data of Irvine and Huebner (87) and on unpublished data for chloralose preparation of D. P. Phillips and D. R. F. Irvine. In each case, N specifies number of neurons on which distribution is based. Modified from Irvine and Huebner (87).

Two further points relating to the thresholds of association cortex neurons are also illustrated by the data of Fig.5.3. In an earlier study of MSA neurons (171) it was concluded that these cells were generally insensitive. As Fig.5.3 illustrates, however, some association cortex neurons, both broadly and sharply tuned, have low thresholds that are comparable to those of primary auditory neurons. The technical and procedural differences responsible for this discrepancy have been discussed elsewhere (87). The second point concerns those neurons that were in fact insensitive to pure tones (e.g., units 17-5, 24-6, 35-3, and 4-2 in Fig.5.3B,C,E,F, respectively). Every neuron of this sort that was also tested with white noise proved to have a low noise threshold comparable to those of cells with sensitive tone responses. This point is illustrated by the noise threshold for units in Fig.5.3. This characteristic was not recognized in the initial analysis of the cortical data. When it was observed in the responses of M/IL thalamus neurons (86; see Fig. 8), however, the cortical data were reexamined and were found to exhibit the same feature. It suggests that one consequence of the considerable degree of frequency convergence in the input to association cortex neurons is that some of these neurons are more sensitive to spectrally complex sounds than to pure tones. This suggestion is supported by the observation that some units with normal noise thresholds were either not driven at all by pure tones or were driven so weakly that data on tuning could not be obtained.

3.3. Binaural Properties

The final aspect of acoustic input to be considered concerns the binaural properties of association cortex neurons. The overwhelming majority of neurons in all three association areas exhibited an excitatory onset response to monaural stimulation of either ear. In the terminology that has been used to describe laterality in the primary auditory pathway, these cells have been designated E/E (66). It should be emphasized that this designation does not imply exclusively excitatory input from each ear. As the discussion of response patterns here and in later sections indicates, the input to most cells has inhibitory as well as excitatory components. Data on laterality of input are presented in Table 5.1, in which the predominance of the E/E pattern in all three asociation fields is apparent. Control recordings from A I neurons make it clear that this dominance is not attributable to a generalized effect of chloralose anesthesia.

Binaural interaction in E/E cells has been evaluated by means of the "summation ratio" proposed by Goldberg and Brown (67), which expresses the binaural response as a ratio of the sum of the monaural responses. The forms of interaction termed facilitation, summation and occlusion, are defined by ratios greater than, equal

Table 5.1
Binaural Input To Association
Cortex Neurons[a]

	Binaural input pattern			
Area	E/E	E/O or O/E	O/O(E)	Total
ALA	39	1	0	40
PCA	46	1	2	49
MSA	46	2	1	49
AI	8	15	2	25

[a]Number of neurons in association areas and in A I showing different patterns of input from the two ears. E/E neurons exhibit an excitatory response to monaural stimulation of either ear. E/O and O/E cells are excited by monaural stimulation of the contralateral or ipsilateral ear, respectively, but show no response to monaural stimulation of the other ear. O/O(E) neurons show no response to monaural stimulation of either ear but are strongly driven by simultaneous binaural stimuli. Data for A I in chloralose-anesthetized cats are unpublished observations by D. P. Phillips and D. R. F. Irvine. Reproduced from Irvine and Huebner (87).

to and less than unity, respectively. The great majority of neurons in all three association areas are characterized by occlusive interaction (Table 5.2). As discussed in detail elsewhere (86, 87), occlusion reflects the fact that monaural and binaural intensity functions for these neurons become asymptotic at the same levels.

The data summarized in this section provide a more detailed characterization of the acoustic response properties of association

Table 5.2
Binaural Interaction Patterns In Association Cortex[a]

	Binaural interaction pattern			
Area	Occlusion	Summation	Facilitation	Total
ALA	14	3	5	22
PCA	22	1	2	25
MSA	10	3	3	16
Total	46	7	10	63

[a]Number of E/E cells in each association area showing various forms of binaural interaction at 65–75 dB SPL. Terms defined in text. From Irvine and Huebner (87).

cortex single neurons than has been provided by previous studies, most of which were restricted to click stimulation (18, 34, 49, 52, 139). Comparison of these characteristics with those of single neurons in areas considered to be possible sources of acoustic input to the association fields enables a preliminary analysis of the manner in which this input is organized. In the following sections the major sources of input indicated by available anatomical and neurophysiological evidence are examined.

4. The Nonspecific Projection System

4.1. Acoustic Input to Medial/Intralaminar Thalamus

The nonspecific projection system hypothesis predicts homogeneity of acoustic input to the three association fields. The observed identity of acoustic properties in these fields is in accordance with this prediction. According to the hypothesis, these properties should be identical to those in the nonspecific projection system itself. Previous evidence on acoustic input to M/IL thalamus has established only the presence of click-evoked gross responses in this region (10, 37, 111, 157). In this section, recent evidence (86) on the detailed acoustic properties of M/IL thalamic neurons will be summarized and a number of issues raised by the nonspecific projection system hypothesis will be examined. The procedures were similar in most respects to those employed in the study of association cortex and have been described in detail in a previous report (86).

The majority of neurons in these experiments were isolated in the dorsomedial nucleus (MD), or in the center median/parafascicular (CM/Pf) complex, the latter comprising the center median (CM), parafascicular (Pf) and subparafascicular (SPf) nuclei. A representative electrode track through MD and CM is shown in Fig. 5.5. Neurons in both areas were activated by acoustic stimulation (those marked by A in Fig. 5.5) and many acoustically responsive neurons were also driven by visual and/or somatic stimulation (V and S, respectively). Data on acoustic responsiveness and polysensory convergence for the total sample are shown in Fig. 5.6A,B respectively. Comparison with Fig. 5.2 indicates that the proportion of acoustically activated neurons in CM/Pf is similar to that in association cortex, but that the degree of polysensory convergence is considerably less than that observed in cortex under identical test conditions.

The most commonly observed response pattern to acoustic stimulation was an onset response (e.g., Fig. 5.7A), followed—in

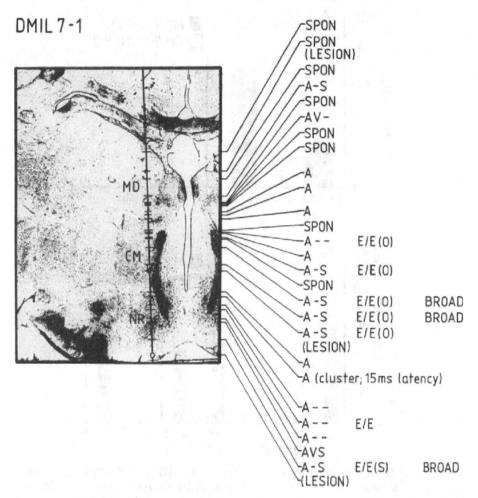

FIG. 5.5. Photomicrograph of near-frontal section through postero-medial thalamus of cat DMIL 7 showing position of electrode track. Locations of lesions are indicated by traced outlines and those of isolated single neurons by horizontal bars. Information in first column at right summarizes sensory input: AVS indicates responses to auditory, visual and somatosensory stimulation; dash in position correspondng to V or S indicates no response to that modality; A alone indicates that only acoustic responsiveness was assessed; SPON indicates that unit was spontaneously active but not responsive to any of the available stimuli. Information in second and third columns describes binaural input and tuning characteristics, respectively, of those units for which they could be determined. Terms defined in text. MD, dorsomedial nucleus; CM, center median nucleus; NR, red nucleus. From Irvine (86).

units with spontaneous activity—by a long-lasting suppression of background activity (e.g., Fig. 5.7B). Two other response patterns were observed in MD. "Complex" responses consisted of an onset

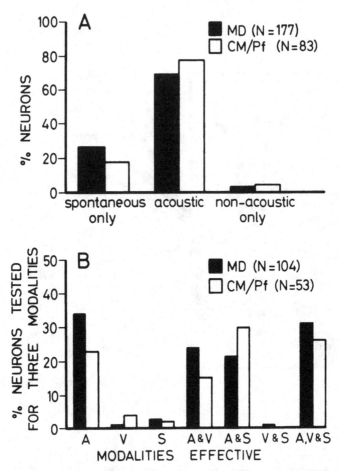

FIG. 5.6. A, proportions of neurons in MD and CM/Pf that were driven acoustically (ACOUSTIC), driven only by non-acoustic stimulation (NONACOUSTIC), or that were not driven by any of the available stimuli (SPONTANEOUS ONLY). One MD neuron was driven only by the simultaneous presentation of auditory and visual stimulation and has not been included. B, proportions of those neurons tested with all three modalities that responded to auditory (A), visual (V) and somatosensory (S) stimulation. Categories are exclusive and conjunctions indicate responses to each of the modalities presented separately. In each case N specifies the number of neurons on which the proportions are based. From Irvine (86).

response followed by suppression of spontaneous activity, and a later discharge (e.g., Fig. 5.7C, D), sometimes of a reverberatory nature (e.g., Fig. 5.7F). "Long latency" responses were characterized by latencies of 300–600 ms (e.g., Fig. 5.7I). Cells of these types exhibited the same temporal discharge pattern in response to short noise bursts (e.g., Fig. 5.7E) or to punctate visual and somatic stimulation (e.g., Fig. 5.7G, H, J). The number of cells exhibiting

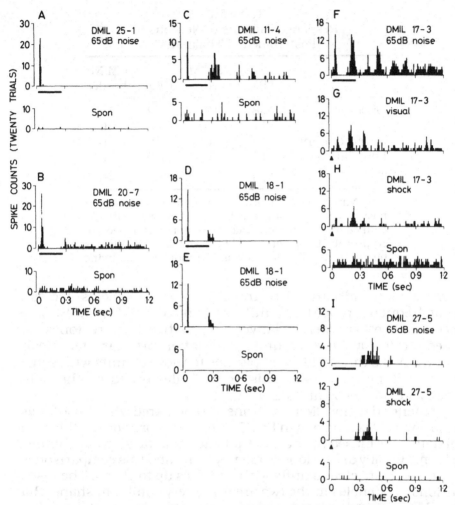

FIG. 5.7. Response histograms for selected units in posteromedial thalamus illustrating response patterns described in text. Unit identification (cat number–unit number) and stimulus parameters are specified in labels. Bottom histogram in each set is for spontaneous (Spon) activity. All histograms are based on 20 trials. Duration of acoustic stimuli is indicated by thick bar beneath abscissa; point at which visual and somatosensory (shock) stimuli were presented is indicated by ▲. From Irvine (86).

these responses in MD and CM/Pf and in their border regions is shown in Table 5.3. It is apparent that response patterns other than onset were almost entirely restricted to MD.

Representative tuning curves for neurons in CM/Pf and MD are presented in Fig. 5.8. As illustrated by this figure, the overwhelming majority of units in both regions had broad, irregular tuning curves of the type observed in association cortex. Although a

Table 5.3
Response Patterns of Neurons In
Posteromedial Thalamus[a]

	MD	MD/CM border	CM/Pf	CM/NR border
Onset	88	19	62	16
Late	11	0	3	0
Reverberatory	5	0	0	0
Long-latency	15	0	0	0
Total	119	19	65	16

[a]Number of neurons in different areas of posteromedial thalamus exhibiting response patterns defined in text. Neurons that could not be assigned unequivocally to MD or to CM/Pf have been classified as border units; NR: red nucleus. Reproduced from Irvine (86).

number of broadly tuned neurons had low thresholds across a broad frequency range (e.g., units 19-4 and 30-5 in Fig. 5.8B, C, respectively), others were relatively insensitive to pure tones. As noted previously, however, units of the latter sort characteristically had low noise thresholds, comparable to those of units with sensitive pure-tone responses. This point is illustrated by the noise thresholds for these units in Fig. 5.8.

Latency data for CM/Pf neurons are compared with those for association cortex neurons in Fig. 5.9. As noted previously, the cortical data were biased by the fact that latency data were not obtained for some weakly driven longer-latency units and this comparison is, therefore, restricted to units with latencies up to 40 ms. The distributions of latencies in the two areas are very similar in shape, that for CM/Pf being displaced toward lower values by about 4 ms. The shapes of the distributions were compared by subtracting 4 ms from all cortical latencies and collapsing the data into 4-ms categories in order to satisfy the requirements of the χ^2 test with respect to low expected frequencies. The shapes of the adjusted distributions were not significantly different ($\chi^2 = 2.88$; $df = 4$; $p > 0.5$).

Information on the binaural properties of neurons in CM/Pf and in MD are summarized in Tables 5.4 and 5.5. As for the association cortex, the majority of units were classified as E/E (Table 5.4), and most of these units were characterized by occlusive binaural interaction (Table 5.5). Binaural interaction patterns in E/E cells were stable except at near-threshold levels.

It is clear from the preceding account that there is almost perfect identity between the acoustic response properties of neurons in CM/Pf and in the cortical association fields. That this is not a

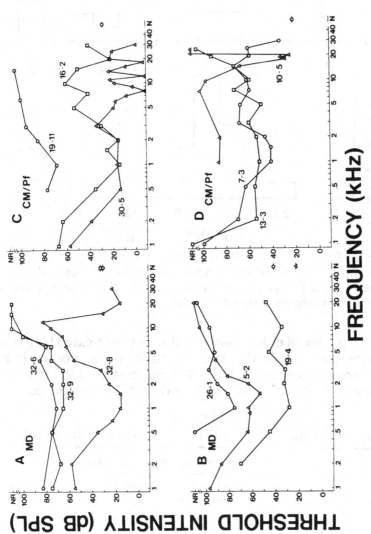

FIG. 5.8. Tuning curves for representative neurons in the thalamic regions indicated in the upper left-hand corner of each graph. Conventions as in Fig. 5.3. From Irvine (86).

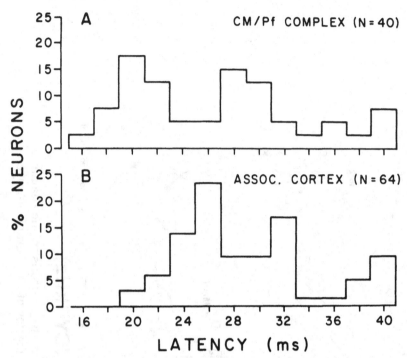

FIG. 5.9. Latency distributions for neurons with latency up to 40 ms in CM/Pf complex and in the cortical association areas. Cortical data are unpublished results from Irvine and Huebner (87). N indicates number of neurons in each category. From Irvine (86).

trivial consequence of chloralose anesthesia is indicated by the fact that response patterns in MD, and many acoustic response characteristics of areas examined in the following section, differ in a number of respects from those of CM/Pf and association cortex.

Table 5.4
Binaural Input To Neurons In Posteromedial Thalamus[a]

Area	Binaural input pattern				
	E/E	E/O or O/E	O/O(E)	Unclear	Total
MD	33	1	10	3	47
MD/CM border	4	1	2	1	8
CM/Pf	37	2	1	0	40
CM/NR border	7	1	0	0	8
Total	81	5	13	4	103

[a]Number of onset cells in each region showing different patterns of input from the two ears. For details, see text and footnote to Table 5.1. Reproduced from Irvine (86).

Table 5.5
Binaural Interaction Patterns In Posteromedial Thalamus[a]

Area	Binaural interaction pattern			
	Occlusion	Summation	Facilitation	Total
MD	17	5	7	29
MD/CM border	2	2	2	6
CM/Pf	24	4	7	35
CM/NR border	4	1	1	6
Total	47	12	17	76

[a]Number of E/E onset cells in each region showing various forms of binaural interaction to 65 dB noise-burst stimulation. Terms defined in text. Reproduced from Irvine (86).

Both this homogeneity of response properties and the latency relationship between CM/Pf and association cortex are in accordance with the nonspecific projection system hypothesis. These data, therefore, provide indirect support for the view that acoustic input to the cortical association fields is derived from M/IL thalamus.

4.2. Cortical Projections from Medial/Intralaminar Thalamus

A major difficulty for the view that the intralaminar nuclei are involved in the transmission of sensory information to association cortex has been the uncertainty concerning the existence of cortical projections from this region. The long-standing controversy about the existence of *direct* cortical projections (30, 117, 130) has been reviewed by others (93, 99, 109) and was attributable to the difficulty in interpreting degeneration in these nuclei following cortical lesions. More recently, use of the horseradish peroxidase (HRP) tracing technique (88, 89, 93, 99, 108, 165) has confirmed electrophysiological evidence (9, 16, 55, 132) for the existence of such projections in addition to the well-established projection to the striatum. However, there remains uncertainty concerning the manner in which the projections are organized. Jones and Leavitt (93) concluded that the projection of the intralaminar nuclei (including CM and Pf) upon the cortex in cat, rat and monkey was sparse and diffuse. However, Kennedy and Baleydier (99) have recently demonstrated a considerable degree of specificity of projections from intralaminar thalamus to visual cortex in the cat: Labeled cells were found in intralaminar nuclei, particularly the central lateral nucleus (CL), following injections in areas 18, 19 and the Clare-Bishop area, but not after injections in area 17. Thus, although the existence of a nonspecific thalamocortical projection system is confirmed by these studies, its organization remains unclear. Furthermore, little is known about its relation to the cortical association fields. There is electrophysiological and anatomical evidence for projections from CL and CM/Pf to pericruciate cortex and to the anterior lateral and medial suprasylvian gyri (9, 16, 30, 55, 79, 89, 108, 114, 132, 137, 138, 165). There is also evidence, however, for projections from the ventroanterior nucleus (VA) to medial suprasylvian (137, 138) and pericruciate (16, 55) areas, and Robertson and Thompson have suggested that the intralaminar nuclei might project *indirectly* to the cortex by way of VA (141).

The results described in the preceding section do not enable any resolution of these issues. In the course of these experiments, a few neurons were isolated in CL and most were acoustically responsive. However, the sample size was too small to permit characterization of their response properties or comparison with those in CM/Pf. The latency difference of approximately 4 ms between CM/Pf and association cortex suggested by the present data could reflect either a slow direct pathway or the indirect projection proposed by Robertson and Thompson (141). Detailed evidence on the acoustic properties of neurons in CL and VA, and on the pattern of thalamic labeling after HRP injections restricted to the physiologically defined association areas, should help to resolve these issues and clarify the status of the nonspecific projection system hypothesis.

4.3. The Mesencephalic Reticular Formation: Acoustic Input and Connections

The evidence presented here on the acoustic properties of neurons in CM/Pf complex also bears on the organization of acoustic input to this region of thalamus and, thus, on a further aspect of the nonspecific system hypothesis. The CM/Pf complex receives afferents from multiple ascending and descending projection systems. There is both anatomical (31, 54, 75, 118, 148, 151, 152) and electrophysiological (32, 46, 62, 140) evidence for a major projection to CM/Pf complex via the dorsal "leaf" of ascending fibers from the brain stem RF. There are also direct projections from the mesencephalic tectum and pretectum to this region (151) and descending fibers from the thalamic reticular nucleus (92, 149) and from cerebral cortex (134, 135) terminate in CM/Pf complex.

There have been numerous reports of polysensory convergence on neurons in the cat's RF (13, 17, 81, 110, 152). Recent evidence (90; Irvine and Jackson, submitted for publication) on the acoustic properties of neurons in mesencephalic RF indicates a high degree of similarity to those in CM/Pf with respect to frequency tuning and binaural input. The majority of neurons also exhibited simple onset responses. However, a significant proportion exhibited complex response patterns of the type observed in MD and occasional cells exhibited long latency responses. Onset-component modal latencies for the great majority of neurons fell within the range from 12 to 26 ms. These characteristics are in good agreement with the proposal that the major source of acoustic input to intralaminar thalamus is provided by ascending projections from RF, although the absence of complex response patterns in CM/Pf suggests the possibility that this input might be derived from only a subset of RF neurons.

The source of acoustic input to RF is not known. Tecto-reticular projections seem the most likely candidate, but the relative contributions of the inferior and superior colliculi (IC and SC) are unclear. Powell and Hatton (133) reported projections to MRF from IC, and recent autoradiographic tracing studies (Aitkin and Kenyon; personal communication) have indicated a projection to RF from the external nucleus of the IC (ICX), neurons in which exhibit broad tuning and polysensory convergence (5, 7). There is also, however, strong evidence for projections from the deep layers of SC to mesencephalic and ponto-medullary RF (12, 70, 97, 98). Many neurons in deep SC are acoustically responsive (e.g., 69, 155, 173), deriving their input in part from IC (53, 131), specifically ICX and ICP (53). Some further implications of these data will be considered in subsequent sections.

4.4. Conclusions and Commentary

One feature of acoustic input to the nonspecific system that deserves further consideration is the high degree of convergence reflected in the tuning and binaural properties of neurons throughout the system. Acoustic input is conveyed to the central nervous system by auditory nerve fibers, all of which are sharply tuned, i.e., respond to only a limited frequency range at low SPLs (100). This characteristic is preserved throughout the tonotopically organized core of the auditory lemniscal pathway (4). Broad tuning at similar SPLs must, therefore, reflect convergence onto some single neurons of inputs from a large number of sharply tuned elements whose best frequencies span the spectral range of broadly tuned units. In the case of extremely sensitive broadly tuned units, each of these inputs must individually be capable of exciting the neuron at low SPLs. Other neurons, however, are unresponsive or insensitive to pure tones, but have low thresholds to noise. In neurons of this type, summation of simultaneous input from many frequency-specific sources is apparently necessary to activate the cell at other than high SPLs. This characteristic has the consequence that these neurons are more sensitive to, and in that sense specialized for the detection of, spectrally complex sounds.

The data reviewed in this section have been largely descriptive and correlational. Although this evidence does not constitute a critical test of the nonspecific projection system hypothesis, it provides circumstantial support for the involvement of this system in the transmission of acoustic input to the association fields. It is apparent from the evidence concerning visual input to MSA neurons (47, 48, 50), however, that there might well be multiple sources of sensory input to the association fields. This possibility is also indicated by electrical stimulation evidence for convergence onto ALA and MSA neurons of input from the nonspecific system and from the thalamic pulvinar/lateral posterior complex (52, 101, 153). Before attempting a final evaluation of these data, therefore, evidence concerning other potential sources of acoustic input will be considered in the following section.

5. The Auditory "Lemniscal Adjunct" System

5.1. The "Lemniscal Line"/ "Lemniscal Adjunct" Distinction

The observed homogeneity of acoustic input to the three association fields is also, of course, compatible with their input being derived from some other acoustic region (or regions) having the same characteristics, or with similar convergence occurring independently in each of the three fields. Analysis of the regions that might be involved if the organization were of this sort requires consideration of possible divisions within higher levels of the primary auditory pathway. Graybiel (71, 73, 74) has proposed for all sensory systems a distinction between what she terms "lemniscal line" and "lemniscal adjunct" systems. The former are broadly characterized as core systems that maintain topographic and modality specificity. "Lemniscal adjunct" systems consist of regions adjacent to the lemniscal line that are more heterogeneous and characterized by greater convergence of sensory influences. Graybiel defines the higher levels of the auditory "lemniscal line" system as comprising the central nucleus of the IC (ICC), the ventral division of the medial geniculate (MGv) and primary auditory cortex (presumably A I). The "adjunct" system comprises the external and pericentral nuclei of the colliculus (ICX and ICP, respectively), the thalamic "posterior nuclear group" (PO; 71, 73, 78, 145), which includes the medial division of MG (MGm), and "rim" or "belt" areas surrounding the primary auditory cortical fields.

A good deal of recent neurophysiological evidence supports a functional distinction of the sort proposed by Graybiel. Thus, neurons in ICC, MGv and A I respond only to acoustic stimulation, are securely driven, and exhibit sharp tuning. Within each area, neurons are organized in a precise tonotopic sequence (4, 6, 7, 85, 112, 113, 127, 133, 144). In contrast, many neurons in the designated midbrain and thalamic adjunct areas are labile, exhibit broad irregular tuning and receive polysensory input (3, 5, 7, 27, 106, 126, 170). At the cortical level, the functional distinction is less clear, because of recent evidence for the existence of multiple tonotopically organized fields (85, 102, 112, 133) and the paucity of data on the properties of neurons in the remaining "belt" areas (112, 133; see Imig et al.'s chapter in this volume). More recent neuroanatomical tracing studies, however, have prompted Andersen et al. (14, 15; see Merzenich, Colwell, and Andersen's chapter in this volume) to propose a distinction between "cochleotopic" and "diffuse" auditory systems. Although the thalamocortical relations revealed by these studies are more complex than those assumed by the line–adjunct model, this distinction corresponds rather closely to that proposed by Graybiel.

It is apparent from this brief survey that the acoustic properties of association cortex neurons resemble those of the diffuse "adjunct" system rather than those of the cochleotopic "line" system. Available anatomical evidence also indicates that any input to association cortex from the auditory pathway is likely to be derived from the diffuse or "adjunct" system.

5.2. The Pulvinar–Posterior Complex: Acoustic Input and Connections

One such potential source of acoustic input to the posterior association fields is indicated by evidence on the cortical projections of the thalamic pulvinar–posterior (Pul–PO) complex. There have been a number of reports of projections from Pul to the medial suprasylvian gyrus (40, 72, 172). Graybiel (72) also reported that lesions in the rostral and dorsal part of the lateral posterior nucleus (LP), which she designated lateralis intermedius (LI) after Rioch (136), resulted in degeneration on the crown of the suprasylvian gyrus. Electrophysiological evidence indicates that these projections terminate, at least in part, in the suprasylvian association field (52, 101, 153, 156). Their origins have been defined more precisely by recent HRP tracing experiments, in which suprasylvian injections

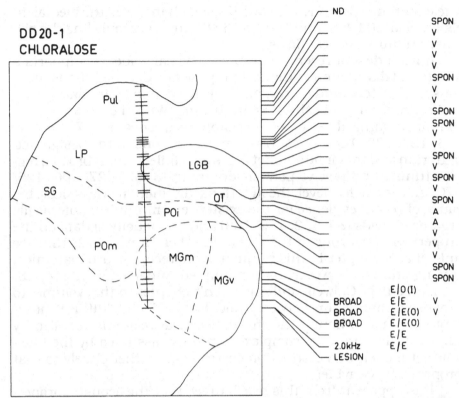

FIG. 5.10 Schematic drawings of electrode penetrations through Pul–PO complex at similar lateral and anterior–posterior coordinates, in two animals differing in anesthesia employed. Location of isolated single neurons indicated by horizontal bars. At right, first column indicates the frequency-tuning characteristics of auditory unit at that location: Broadly tuned units are indicated by BROAD; best frequency is indicated for sharply tuned units. ND denotes loss of driven activity. Second column in-

result in parallel bands of labeled cells in Pul and LI (20; Irvine and Brugge, unpublished observations).

    Graybiel (72) has also reported that some lesions of LP produce degeneration in parts of areas 5 and 7, which appear to correspond to the location of ALA. Both LP (71, 72) and the subdivisions of the posterior group (PO; 71, 73, 78) also project to sulcal "rim" areas lying between the auditory and posterior association fields. On the basis of these observations, Pul and LP have been implicated in the transmission of sensory input to the posterior association areas (71, 74).

    Although there have been a number of reports of acoustic responses in the Pul–PO complex (19, 27, 39, 43, 64, 83, 106, 129, 170), the only previous analysis of specific acoustic properties in

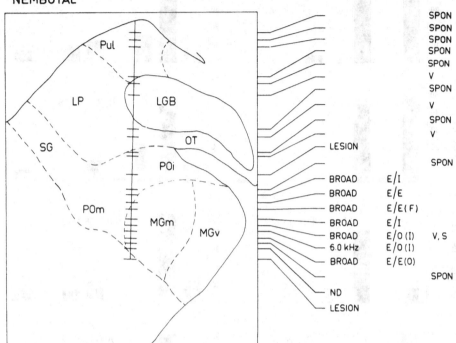

dicates laterality and binaural interaction properties for acoustic cells. Symbols as defined in text. Third column denotes modalities to which unit at that location responds. Abbreviations: SPON, spontaneously active but not driven; V, visually driven; S, somatically driven; A, acoustically driven, but data presented in first two columns not obtained. Histological abbreviations: SG, nucleus suprageniculatus thalami; LGB, lateral geniculate body; OT, optic tract; all others as defined in text. From Phillips and Irvine (126).

this region has been Aitkin's (3) study of MGm. In the remainder of this section, evidence from a recent study of input to neurons in the Pul–PO complex (126) will be summarized. This evidence makes possible an assessment of the contribution of this region to the acoustic input to the association fields.

Schematic drawings of electrode penetrations through the Pul–PO complex are presented in Fig. 5.10. These representative penetrations illustrate two general findings of the study. The first is that, if chloralose and barbiturate preparations are compared, the proportion of neurons activated by sensory stimulation was depressed by barbiturate in Pul, but was not anesthetic-dependent in MGm or the other divisions of PO. The second point is that, whereas the majority of responsive neurons in Pul and LP were acti-

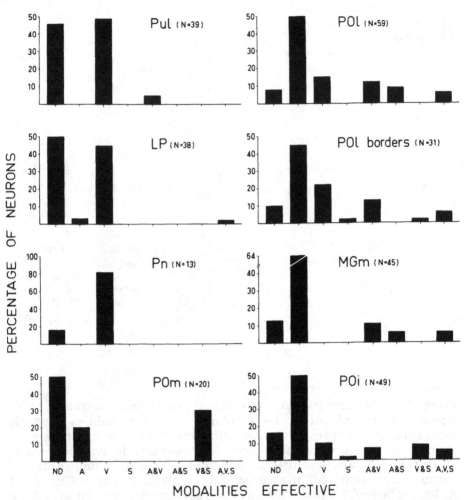

FIG. 5.11. Proportions of cells in various divisions of the Pul–PO complex responding to different stimulus modalities. Categories are exclusive and conjunctions indicate response to each modality presented separately. N indicates sample size for each area. Abbreviations: ND, not driven; A, acoustic; V, visual; S, somatic. For details, see text. From Phillips and Irvine (126).

vated only by visual stimulation, the majority of those in PO were acoustically responsive. Data on sensory input for the entire sample are presented in Fig. 5.11. In these figures, and in the ensuing discussion, the parcellation of PO employed is that proposed by Jones and Powell (94). It is apparent from Fig. 5.11 that sensory responsiveness in Pul and in LP was restricted almost entirely to visual stimulation; the proportion of visually responsive neurons is undoubtedly an underestimate, however, since chloralose and barbiturate data were combined in this compilation. In all except the medial division (POm), the majority (75–85%) of neurons in the various divisions of PO were acoustically responsive. Although some of these neurons also responded to other modalities, comparison of these data with those in Figs. 5.2 and 5.6 indicates that the proportions of polysensory neurons are much smaller than those in the nonspecific system.

The failure to observe acoustically responsive units in Pul and LP is apparently at variance with earlier reports of auditory evoked responses in the region (83, 84, 104). In these earlier studies, however, identification of recording sites was based on a stereotaxic atlas (91) in which LP is so defined as to include most of what has subsequently been identified as PO. Inspection of the acoustically responsive sites ascribed to LP in these studies indicates that many would in fact now be assigned to PO. It is also possible that some responses in Pul might have reflected volume conduction from other parts of the complex.

As indicated previously, Pul and LP are the two parts of the Pul–PO complex known to project to the regions of the posterior cortical association fields. The absence of acoustic input to Pul and LP, however, indicates that these areas are unlikely to be involved in the transmission of acoustic input to the cortex. On the other hand, the strong visual input to neurons in Pul and LP is in accordance with previous reports (64, 65, 166); the detailed visual response properties of these units have been reported (65) to resemble those of MSA neurons (47, 48), and it seems probable that these regions are involved in the transmission of visual input to the posterior association fields (71, 74).

Although strong acoustic drive was characteristic of all but the medial division of PO, there were differences between the divisions with respect to their acoustic properties. These differences bear mainly on the organization of the auditory lemniscal adjunct system and will only be briefly summarized here. Evidence on tuning characteristics is presented in Fig. 5.12. Whereas the majority of units in the lateral division (PO1) were sharply tuned, the majority of those in MGm and in the intermediate division (POi) exhibited broad tuning. Most units in POi and MGm had short latencies (mean modal latencies of 16.2 ms and 15.2 ms, respectively), but many POi neurons had much longer latencies (mean of 34.7 ms). The long latency of most POi cells is in accordance with anatomical evidence that auditory input to this region is derived predominantly from corticothalamic projections (45, 94). Both MGm and PO1 receive substantial projections from the IC (71, 94, 115). The difference in tuning between these divisions of PO, however, suggests that their input might largely be derived from different divisions of IC, that to POi originating in ICC and that to MGm in either ICX or ICP. This suggestion is in accordance with the

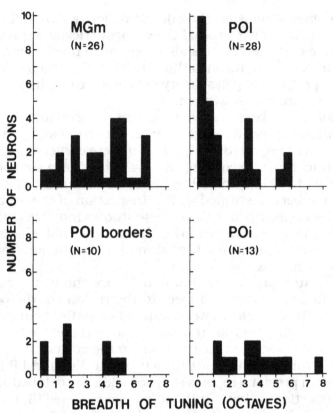

FIG. 5.12. Breadth of tuning distribution for neurons in four divisions of PO. In each case N specifies number of neurons on which distribution is based. Modified from Phillips and Irvine (126).

evidence on the connections of these structures presented by Andersen et al. (14, 15; see Merzenich et al.'s chapter in this volume) and with their suggestion that PO1 should perhaps be considered part of the "cochleotopic" line system.

5.3. Possible Input from Auditory Cortex

The evidence on the acoustic properties of the Pul–PO complex presented here indicates that this region is unlikely to be a direct source of acoustic input to the association fields. As mentioned previously, however, the various divisions of PO project to auditory and periauditory cortex (14, 15, 71, 73, 78) and might, therefore, be related to the association areas indirectly. Evaluation of this possibility is complicated by the fact that the location of the various auditory fields relative to sulcal patterns varies between animals and must be determined by physiological mapping (112, 133). In the only study in which the cortical connections of physiologically defined auditory cortical regions have been examined, no evidence of projections to the suprasylvian or lateral gyri was found after isotope injections restricted to A I or to the anterior (A), posterior (P) and ventral posterior (VP) tonotopic fields (85; see Imig et al.'s chapter in this volume). A number of earlier studies, however, reported terminal degeneration in regions corresponding to the three association fields after lesions in "belt" areas surrounding the primary auditory fields (44, 79, 125). Although auditory cortical fields were not defined physiologically in these earlier studies, the lesions appear to have been located in the caudal part of the area designated suprasylvian fringe (SF) by Woolsey (174), which corresponds to the dorsal posterior field (DP) of Imig and Reale (85). This region receives projections from other parts of auditory cortex (44, 85, 96) and from PO (73, 78). Unfortunately, there has been no detailed account of the acoustic properties of neurons in DP. In cats under barbiturate anesthesia, some DP neurons have been reported to be poorly driven by tones and not to have clear best frequencies (112, 133); others are securely driven and tend to have high best frequencies (82, 133). In awake cats, neurons in that portion of DP located on the lateral wall of the suprasylvian sulcus have been reported to be strongly driven by complex acoustic stimuli (124). Although these anatomical and physiological data are limited, they suggest the possibility that this region of auditory cortex might constitute a source of acoustic input to the cortical association fields. In this context, it is of interest that projections from an analogous rim area—the superior temporal gyrus—have been implicated in acoustic input to the frontal polysensory area in the squirrel monkey (23–25, 61), a region that is probably analogous to PCA in the cat (24, 25). More detailed information on the connections and response characteristics of physiologically defined DP will be needed finally to evaluate this possibility.

6. Chloralose Anesthesia and the Functional Interpretation of the Data

The association areas of cerebral cortex and the other "nonspecific" areas considered in this chapter are unresponsive to sensory stimulation, or greatly depressed, under barbiturate anesthesia (34). Although there have been numerous reports of sensory evoked activity in the association cortex of unanesthetized cats (18, 35, 38, 51, 56, 57, 121, 122, 139, 153, 171), most detailed studies of sensory input have employed chloralose anesthesia, because of the enhancement of evoked activity produced by this anesthetic agent.

Despite the evidence from unanesthetized animals, there is a frequently expressed view that the activity seen under chloralose is abnormal, in the sense that the afferent pathways responsible for this activity might not be functional in the normal animal. It is, therefore, important to determine the extent to which conclusions concerning the organization of acoustic input to the association fields derived from the data presented in this chapter must be qualified by the fact that these data were obtained using chloralose anesthesia. Evidence on the nature of the input changes produced by chloralose will therefore be reviewed and the implications of these data examined in this section.

The most informative evidence on the effects of chloralose is provided by a small number of experiments, in which recordings were made from the same sites in the same animals before and after injection of the anesthetic. Bignall and Singer (26) reported that acoustic and photic evoked potentials in the frontal polysensory cortex of unanesthetized squirrel monkeys were enhanced by chloralose, but that the anesthetic did not "create" responses at sites that were silent before the injection. Wester et al. (171) recorded the activity of a single MSA neuron in each of a small number of paralyzed, locally anesthetized cats before and after intraveneous administration of chloralose. Chloralose eliminated spontaneous activity in these neurons and transformed relatively diffuse, sustained responses to tonal stimuli into clearly defined onset responses. These data are of particular interest in that they establish the occurrence of facilitation in individual neurons and indicate that the differences between preparations are not simply attributable to sampling from different populations under the two conditions. An effect similar to that described by Wester et al. has been reported by Engel (56) on the basis of intracellular recording of click-evoked responses in PCA. In awake cats, the responses were weak and diffuse and were not associated with consistent EPSPs. In chloralose preparations synchronized bursts of spikes were associated with clear and consistent EPSPs.

The critical question posed by these effects in the present context is whether chloralose merely clarifies the input to association cortex neurons or whether it distorts the input to an extent that invalidates the data. The latter conclusion is suggested by Dubner and Rutledge's (51) report that the proportions of responsive and polymodal units in MSA were increased in chloralose cats relative to unanesthetized animals. More recent data, however, is at variance with this report. Robertson et al. (139) reported similar proportions of polysensory units in MSA of unanesthetized and chloralose cats. Similarly, Schechter and Murphy (147) reported no significant difference between the proportions of driven units in the prearcuate polysensory area of unanesthetized and chloralose-anesthetized squirrel monkeys. Robertson et al. (139) suggested that the effect reported by Dubner and Rutledge (51) might have been attributable to the difficulty of discriminating responses in the presence of the high rate of spontaneous activity observed in unanesthetized and lightly anesthetized preparations. It is certainly the case that the temporally diffuse responses in unanesthetized animals can be difficult to detect without the use of cumulative response histograms, and the different proportions reported by Dubner and Rutledge could have reflected the differential detectability of responses in the two preparations. It is also possible, however, that for some neurons the afferent input in the unanesthetized state is not sufficient to generate action potentials and that chloralose either enhances the synaptic input to suprathreshold levels or lowers the threshold of the neuron.

The weight of the evidence reviewed here suggests that any neuron responding to a given (acoustic) stimulus under chloralose would also respond to that stimulus, or at least receive afferent input, in the unanesthetized state. If this is so, conclusions concerning the pattern of connectivity underlying the sensory input to such neurons can validly be derived from data obtained under chloralose anesthesia. The differences in the characteristics of the neuron's responses in the two conditions would, however, appear to place some restraints on the functional interpretation of these responses.

Two further points can be made in this connection. The first derives from the fact that the evidence for acoustic responses being weak and diffuse in the awake animal and greatly facilitated by chloralose is based entirely on responses to tone and click stimulation. The data presented in this chapter indicate that at least some neurons in the nonspecific system are considerably more sensitive to longer duration, spectrally complex signals than to pure tones or transients. A number of lines of evidence suggest that responses to complex stimuli of this sort might, in fact, be clearly defined in the unanesthetized animal. Thus, Engel and Woody (57) reported that neurons in coronal-pericruciate cortex of awake cat showed a much stronger response to a "hiss" than to a click of the same intensity (Fig. 5.13A,B). Similarly, Newman and Lindsley (120) reported that neurons in frontal cortex of awake squirrel monkey are more responsive to "noisy" vocalizations than to pure tones. Preliminary observations also indicate that clearly defined responses to tape-recorded cat vocalizations can be obtained from multiple-unit clusters in MSA of the unanesthetized cat (D.R.F. Irvine and R. F. Thompson; unpublished observations).

The second point also derives from Engel and Woody's (57)

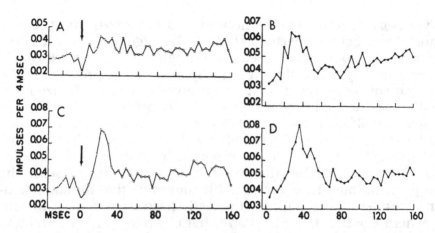

FIG. 5.13. Averaged responses of neurons in coronal-pericruciate cortex of awake cat to click (A and C) and "hiss" (B an D), reproduced from Engel and Woody (57). Click averages are based on 173 neurons and hiss averages on 167 neurons. Ordinate corresponds to number of impulses an average cell of the type studied would have in each 4-ms period across the sampling time. Stimulus was delivered at 0 ms in all cases. Stimuli were free field; hiss was 60-ms duration and the intensity of each stimulus was 60 dB. Click was neutral stimulus in A and conditioned stimulus in C; hiss was neutral in B and conditioned stimulus in D.

study. As Fig. 5.13A illustrates, the response initially evoked by the click stimulus was extremely weak and diffuse. However, when the click acquired significance for the cats as a result of classical conditioning procedures, it elicited clearly-defined synchronous activity (Fig. 5.13C). With respect to the point made in the preceding paragraph, it is of interest that the response to the hiss was not enhanced to nearly the same extent when it was used as a CS (Fig. 5.13B,D). The facilitation of the click response shown here is at least qualitatively similar to that produced by chloralose. The enhancement produced by chloralose is, of course, not selective; it is a general effect on responses to all stimuli. However, this similarity suggests the possibility that the enhancement by chloralose of the response to any particular stimulus might resemble the specific enhancement that occurs in the awake animal when that stimulus acquires significance for the organism. If this were the case, it would suggest a more general interpretation of the high degree of inter- and intramodality convergence exhibited by association cortex neurons. On this view, the convergence seen under chloralose would indicate the *range* of potential sources of afferent influence provided by underlying patterns of connectivity. Which of these afferent inputs, in fact, influenced the cell would be determined by the animal's attentional state and/or the significance of the stimuli impinging on it. A similar view has recently been put forward by

Towe and his colleagues on the basis of the enhancement of sensorimotor cortex unit responses to somatic stimuli under chloralose and during various states of wakefulness (77, 154). Enhancement of this sort by attentional processes has also been reported for visual responses in parietal association cortex (area 7) of awake primates (107, 142).

These considerations concerning the functional interpretation of responses recorded under chloralose are admittedly speculative. They do, however, provide further support for the conclusion that the chloralose preparation can provide valid information concerning the organization of sensory input to the association fields. They also bring out the point that the value of conclusions drawn from "the awake" preparation is also limited, unless consideration is given to the animal's behavioral and attentional state.

7. Conclusions and Final Considerations

The evidence reviewed in this chapter provides indirect support for the view that acoustic input is conveyed to the cortical association fields in the cat via a nonspecific projection system comprising the RF and M/IL thalamus. At various points in the course of the chapter, the need has been indicated for further evidence on certain issues to make possible a final evaluation of this hypothesis. It is of some interest, however, to consider the implications of the existence of such a system in the context of more general views on the organization of the primary auditory pathway.

Recent anatomical and physiological evidence (reviewed in section 5.1) has suggested that the higher levels of the classically defined auditory pathway comprise two largely segregated systems, the cochleotopic "line" system and the diffuse "adjunct" system. If the existence of the nonspecific system is confirmed by further research, it would appear that there are, in fact, three related, but anatomically and physiologically distinct, systems over which auditory information is conveyed to different acoustically responsive regions of the cerebral cortex.

The distinction between the cochleotopic "line" and diffuse "adjunct" systems is first manifested at the colliculus in the differences between the properties and projection patterns of ICC and ICP. Although there have been suggestions of a functional division at lower brain stem levels of the auditory pathway (59, 167, 168), this division does not correspond to that between the line and adjunct systems and there have been no reports of adjunct system properties at auditory centers below the IC. Correspondingly, the mesencephalic RF has been reported to be more responsive to acoustic stimulation than more caudal areas in both the cat (13) and the rat (76). If ICX or deep SC proves to be the major source of acoustic input to RF, as suggested by preliminary evidence discussed in section 4.3., then it would seem that both the adjunct and nonspecific systems might have their origin at the midbrain level.

The existence of three anatomically and physiologically distinguishable systems suggests that they would also be functionally different. However, the functions served by the systems are not known and only the most tentative speculation is possible on the basis of their physiological properties. Neurons in the cochleotopic "line" system are sharply tuned and many are exquisitely sensitive to the intensity- and time-difference cues for the location of a sound source (e.g., 28, 29, 58, 63, 67, 143, 164). This system, thus, has the capability for analysis of the spectral and spatial attributes of acoustic signals and seems likely to carry out at least these functions. Little is known about the physiological properties of neurons in the diffuse "adjunct" system, other than that they are generally broadly tuned and do not appear to be tonotopically organized. It is of some interest in this context that Knudsen and Konishi (103) have shown that different regions in an auditory midbrain nucleus of the owl, which is homologous to the mammalian IC, have functionally distinct organizations. One region is nontonotopic and contains a systematic map of auditory space that is independent of spectral properties. The other region is tonotopically, but not spatially, organized. This observation has prompted speculation that the cat's diffuse adjunct system might also be concerned with the representation of acoustic space (14).

Evidence reviewed in the previous section suggests that the nonspecific system is concerned less with the physical attributes of acoustic signals than with their meaning or significance, i.e., with factors relating to the attentional state of the animal. A considerable body of clinical and experimental evidence has established the involvement in attentional processes of parietal association areas in humans and in subhuman primates (e.g., 42, 80, 107, 116, 123, 142, 169). This evidence has mainly concerned visual spatial attention, but there have been some reports of auditory attention deficits associated with parietal and frontal lesions in monkeys (80, 169). Unfortunately, there is no evidence on the effects of association cortex lesions on auditory attention in cats and evidence on the effects of such lesions on auditory discrimination tasks is scant and equivocal (41, 158, 161). The paucity of data on these issues emphasizes the tentative nature of these speculations concerning the functional significance of acoustic input to the cat's association cortex.

Acknowledgments

The research reported in this chapter was supported in part by grants from the Australian Research Grants Committee.

The chapter was prepared while the first author was on study leave in the Department of Neurophysiology of the University of Wisconsin. I am grateful to the members of the Department, who helped to make this time both valuable and enjoyable, and to Dr. Clinton N. Woolsey for the opportunity to participate in the Dallas symposium.

The photographic assistance of Shirley Hunsaker and the secretarial assistance of Evadine Olsen and Vicky Lockwood in the preparation of the manuscript are gratefully acknowledged.

References

1. ABELES, M., AND GOLDSTEIN, M. H., JR. Functional architecture in cat primary auditory cortex: columnar organization and organization according to depth. *J. Neurophysiol.*, 33: 172–187, 1970.
2. ADRIAN, H. D., GOLDBERG, J. M., AND BRUGGE, J. F. Auditory evoked cortical potentials after lesions of brachium of inferior colliculus. *J. Neurophysiol.*, 29: 456–466, 1966.
3. AITKIN, L. M. Medial geniculate body of the cat: responses to tonal stimuli of neurons in medial division. *J. Neurophysiol.*, 36: 275–283, 1973.
4. AITKIN, L. M. Tonotopic organization at higher levels of the auditory pathway. In: *International Review of Neurophysiology,* Vol. 10. *Neurophysiology II,* edited by R. PORTER. Baltimore: University Park Press, 1976, pp. 249–279.
5. AITKIN, L. M., DICKHAUS, H., SCHULT, W., AND ZIMMERMANN, M. External nucleus of inferior colliculus: auditory and spinal somatosensory afferents and their interactions. *J. Neurophysiol.*, 41: 837–847, 1978.
6. AITKIN, L. M., AND WEBSTER, W. R. Medial geniculate body of the cat: Organization and responses to tonal stimuli of neurons in ventral division. *J. Neurophysiol.*, 35: 365–380, 1972.
7. AITKIN, L. M., WEBSTER, W. R., VEALE, J. L., AND CROSBY, D. C. Inferior colliculus. I. Comparison of response properties of neurons in central, pericentral, and external nuclei of adult cat. *J. Neurophysiol.*, 38: 1196–1207, 1975.
8. ALBE-FESSARD, D., AND FESSARD, A. Thalamic integrations and their consequences at the telencephalic level. *Prog. Brain Res.*, 1: 115–148, 1963.
9. ALBE-FESSARD, D., LEVANTE, A., AND ROKYTA, R. Cortical projections of cat medial thalamic cells. *Int. J. Neurosci.*, 1: 327–338, 1971.

10. ALBE-FESSARD, D., AND MALLART, A. Existence de réponses d'origines visuelle et auditive dans le centre médian du thalamus du chat anesthésié au chloralose. *Compt. Rend. Acad. Sci., Paris,* 251: 1040–1042, 1960.

11. ALBE-FESSARD, D., AND ROUGEUL, A. Activités d'origine somesthésique évoquées sur le cortex non-spécifique du chat anesthésié au chloralose: Rôle du centre médian du thalamus. *Electroencephalog. Clin. Neurophysiol.,* 10: 131–152, 1958.

12. ALTMAN, J., AND CARPENTER, M. B. Fiber projections of the superior colliculus in the cat. *J. Comp. Neurol.* 116: 157–178, 1961.

13. AMASSIAN, V. E., AND DEVITO, R. Unit activity in reticular formation and nearby structures. *J. Neurophysiol.,* 17: 575–603, 1954.

14. ANDERSEN, R. A. *Patterns of connectivity of the auditory forebrain of the cat.* PhD Thesis, Uiversity of California, San Francisco, 1979.

15. ANDERSEN, R. A., KNIGHT, P. L., AND MERZENICH, M. M. The thalamocortical and corticothalamic connections of A I, A II, and the anterior auditory field (AAF) in the cat: Evidence for two largely segregated systems of connections. *J. Comp. Neurol.* 194: 649–662, 1980.

16. ARAKI, T., AND ENDO, K. Short latency EPSPs of pyramidal tract cells evoked by stimulation of the centrum medianum-parafascicular complex and the nucleus ventralis anterior of the thalamus. *Brain Res.,* 113: 405–410, 1976.

17. BELL, E., SIERRA, G., BUENDIO, N., AND SEGUNDO, J. P. Sensory properties of neurons in the mesencephalic reticular formation. *J. Neurophysiol.,* 27: 961–987, 1964.

18. BENTAL, E., AND BIHARI, B. Evoked activity of single neurons in sensory association cortex of the cat. *J. Neurophysiol.,* 26: 207–214, 1963.

19. BERKLEY, K. Response properties of cells in ventrobasal and posterior group nuclei of the cat. *J. Neurophysiol.,* 39: 940–952, 1973.

20. BERSON, D. M., AND GRAYBIEL, A. M. Parallel thalamic zones in the LP-pulvinar complex of the cat identified by their afferent and efferent connections. *Brain Res.,* 147: 139–148, 1978.

21. BETTINGER, L. A., DAVIS, J. L., MEIKLE, M. B., BIRCH, H., KOPP, R., SMITH, H. E., AND THOMPSON, R. F. "Novelty" cells in association cortex of cat. *Psychon. Sci.,* 9: 421–422, 1967.

22. BIGNALL, K. E. Effects of subcortical ablations on polysensory cortical responses and interactions in the cat. *Exptl. Neurol.,* 18: 56–67, 1976.

23. BIGNALL, K. E. Bilateral temperofrontal projections in the squirrel monkey: origin, distribution and pathways. *Brain Res.,* 13: 319–327, 1969.

24. BIGNALL, K. E. Auditory input to frontal polysensory cortex of the squirrel monkey: possible pathways. *Brain Res.,* 19: 77–86, 1970.

25. BIGNALL, K. E., AND IMBERT, M. Polysensory and corticocortical projections to frontal lobe of squirrel and rhesus monkeys. *Electroencephalog. Clin. Neurophysiol.,* 26: 206–215, 1969.

26. BIGNAL. K. E, AND SINGER, P. Auditory, somatic and visual input to association and motor cortex of the squirrel monkey. *Exptl. Neurol.,* 18: 300–312, 1967.

27. BLUM, P. S., ABRAHAM, L. D., AND GILMAN, S. Vestibular, auditory and somatic input to the posterior thalamus of the cat. *Exptl. Brain Res.* 34: 1–9, 1979.

28. BOUDREAU, J. C. AND TSUCHITANI, C. Binaural interaction in the cat superior olive S-segment. *J. Neurophysiol.,* 31: 442–454, 1968.

29. BOUDREAU, J. C., AND TSUCHITANI, C. Cat superior olive S-segment cell discharge to tonal stimulation. In: *Contributions to Sensory Physiology,* edited by W. D. NEFF. New York: Academic Press, 4: 143–213, 1970.

30. BOWSHER, D. Some afferent and efferent connections of the parafascicular-center median complex. In: *The Thalamus,* edited by D. P. PURPURA AND M. D. YAHR. New York: Columbia University Press, 1966, pp. 99–108.

31. BOWSHER, D. Diencephalic projections from the midbrain reticular formation. *Brain Res.,* 95: 211–220, 1975.

32. BOWSHER, D., MALLART, A., PETIT, D., AND ALBE-FESSARD, D. A bulbar relay to the centre median. *J. Neurophysiol.,* 31: 288–300, 1968.

33. BRUNER, J. Afférences visuelles non-primaires vers le cortex cérébral chez le chat. *J. Physiol., Paris,* 57, Suppl 12: 1–129, 1965.

34. BUSER, P., AND BIGNALLL, K. E. Nonprimary sensory projections on the cat neocortex. *Int. Rev. Neurobiol.,* 10: 111–165,1967.

35. BUSER, P., AND BORENSTEIN, P. Réponses somesthésiques, visuelles et auditives, recueillies au niveau du cortex "associatif" suprasylvien chez le chat curarisé non anesthésié. *Electroencephalog. Clin. Neurophysiol.,* 11: 285–304, 1959.

36. BUSER, P., BORENSTEIN, P., AND BRUNER, J. Étude des systèmes "associatifs" visuels et auditifs chez le chat anesthésié au chloralose. *Electroencephalog. Clin. Neurophysiol.,* 11: 305–324, 1959.

37. BUSER, P., AND BRUNER, J. Réponses visuelles et acoustiques au niveau du complexe ventromédian posterieur du thalamus chez le chat. *Compt. Rend Acad. Sci., Paris.* 251: 1238–1240, 1960.

38. BUSER, P., AND IMBERT, M. Sensory projections to the motor cortex in cats: a microelectrode study. In: *Sensory Communication,* edited by W. A. ROSENBLITH. Cambridge: MIT Press, 1961, pp. 607–626.

39. CALMA, I. The activity of the posterior group of thalamic nuclei in the cat. *J. Physiol., London,* 180: 350–370, 1965.

40. CLÜVER, P. F. DE V., AND CAMPOS-ORTEGA, J. A. The cortical projection of the pulvinar in the cat. *J. Comp. Neurol.,* 137: 295–308, 1969.

41. CRANFORD, J. L. Polysensory cortex lesions and auditory frequency discrimination in the cat. *Brain Res.,* 148: 499–503, 1978.

42. CRITCHLEY, M. *The Parietal Lobes.* London: Arnold, 1953.

43. CURRY, M. J. The effects of stimulating somatic sensory cortex on single neurons in the posterior group (PO) of the cat. *Brain Res.*, 44: 463–481, 1972.

44. DIAMOND, I. T., JONES, E. G., AND POWELL, T. P. S. The association connections of the auditory cortex of the cat. *Brain Res.*, 11: 560–579, 1968.

45. DIAMOND, I. T., JONES, E. G., AND POWELL, T. P. S. The projection of the auditory cortex upon the diencephalon and brainstem in the cat. *Brain Res.*, 15: 305–340, 1969.

46. DILA, C. J. A midbrain projection to the centre median nucleus of the thalamus. A neurophysiological study. *Brain Res.*, 25: 63–74, 1971.

47. DOW, B. M., AND DUBNER, R. Visual receptive fields and responses to movement in an association area of cat cerebral cortex. *J. Neurophysiol.*, 32: 773–784, 1969.

48. DOW, B. M., AND DUBNER, R. Single unit responses to moving visual stimuli in middle suprasylvian gyrus of the cat. *J. Neurophysiol.*, 34: 47–55, 1971.

49. DUBNER, R. Single cell analysis of sensory interaction in anterior lateral and suprasylvian gyri of the cat cerebral cortex. *Exptl. Neurol.*, 15;: 255–273, 1966.

50. DUBNER, R., AND BROWN, F. J. Responses of cells to restricted visual stimuli in an association area of cat cerebral cortex. *Exptl. Neurol.*, 20: 70–86, 1968.

51. DUBNER, R., AND RUTLEDGE, L. T. Recording and anaylsis of converging input upon neurons in cat association cortex. *J. Neurophysiol.*, 27: 620–634, 1964.

52. DUBNER, R., AND RUTLEDGE, L. T. Intracellular recording of the convergence of input upon neurons in cat association cortex. *Exptl. Neurol.*, 12: 349–369, 1965.

53. EDWARDS, S. B., GINSBURGH, C. L., HENKEL, C. K., AND STEIN, B. E. Sources of subcortical projections to the superior colliculus in the cat. *J. Comp. Neurol.* 184: 309–329, 1979.

54. EDWARDS, S. B., AND DE OLMOS, J. S. Autoradiographic studies of the projections of the midbrain reticular formation: Ascending projections of nucleus cuneiformis. *J. Comp. Neurol.*, 165: 417–432, 1976.

55. ENDO, K., ARAKI, T., AND ITO, K. Short latency EPSPs and incrementing PSPs of pyramidal tract cells evoked by stimulation of the nucleus centralis lateralis of the thalamus. *Brain Res.*, 132: 541–546, 1977.

56. ENGEL, J., JR. Intracellular study of auditory evoked activity in pericruciate cortex of the awake, non-paralyzed cat. *Brain Res.*, 85: 69–73, 1975.

57. ENGEL, J. JR., AND WOODY, C. D. Effects of character and significance of stimulus on unit activity at coronal-pericruciate cortex of cat during performance of conditioned motor response. *J. Neurophysiol.*, 35: 220–229, 1972.

58. ERULKAR, S. D. Comparative aspects of spatial localization of sound. *Physiol. Rev.*, 52: 237–360, 1972.

59. EVANS, E. F. Neural processes for the detection of acoustic patterns and for sound localization. In: *The Neurosciences-Third Study Program*, edited by F. O. SCHMITT AND F. G. WORDEN. Cambridge: MIT Press, 1974, pp. 131–145.

60. EVANS, E. F., AND WHITFIELD, I. C. Classification of unit responses in the auditory cortex of the unanesthetized, unrestrained cat. *J. Physiol., London*, 181: 476–493, 164.

61. FORBES, B. F., AND MOSKOWITZ, N. Cortico-cortical connections of the superior temporal gyrus in the squirrel monkey. *Brain Res.*, 136: 547–552, 1977.

62. FULLER, J. H. Brain stem reticular units: some properties of the course and origin of the ascending trajectory. *Brain Res.*, 83: 349–367, 1975.

63. GEISLER, C. D., RHODE, W. S., AND HAZELTON, D. W. Responses of inferior colliculus neurons in the cat to binaural acoustic stimuli having wide band spectra. *J. Neurophysiol.*, 32: 960–974, 1969.

64. GIULIANO, A., SPREAFICO, R., BROGGI, G., GIOVANNINI, P., AND FRANCESCHETTI, S. Topographic distribution of visual and somesthesic unitary responses in the Pul–LP complex of the cat. *Neurosci. Lett.* 4: 135–143, 1977.

65. GODFRAIND, J.-M., MEULDERS, M., AND VERAART, C. Visual properties of neurons in pulvinar, nucleus lateralis posterior and nucleus suprageniculatus thalami in the cat. I. Qualitative investigation. *Brain Res.*, 44: 503–526, 1972.

66. GOLDBERG, J. M., AND BROWN, P. B. Functional organization of the dog superior olivary complex: An anatomical and electrophysiological study. *J. Neurophysiol.*, 31: 639–656, 1968.

67. GOLDBERG, J. M., AND BROWN, P. B. Responses of binaural neurons of dog superior olivary complex to dichotic tonal stimuli: some physiological mechanisms of sound localization. *J. Neurophysiol.*, 32: 613–636, 1969.

68. GOLDSTEIN, M. H., JR., ABELES, M., DALY, R. L., AND McINTOSH, J. Functional architecture in cat primary auditory cortex: tonotopic organization. *J. Neurophysiol.*, 33: 188–197, 1960.

69. GORDON, B. Receptive fields in deep layers of cat superior colliculus. *J. Neurophysiol.* 36: 157–178, 1973.

70. GRAHAM, J. An autoradiographic study of the efferent connections of the superior colliculus in the cat. *J. Comp. Neurol.* 173: 629–654, 1977.

71. GRAYBIEL, A. M. Some fiber pathways related to the posterior thalamic region in the cat. *Brain, Behav. Evol.*, 6: 363–393, 1972.

72. GRAYBIEL, A. M. Some ascending connections of the pulvinar and lateralis posterior of the thalamus in the cat. *Brain Res.*, 44: 99–125, 1972.

73. GRAYBIEL, A. M. The thalamo-cortical projection of the so-called

posterior nuclear group: a study with anterograde degeneration methods in the cat. *Brain Res.*, 49: 229–244, 1973.

74. GRAYBIEL, A. M. Studies on the anatomical organization of posterior association cortex. In: *The Neurosciences-Third Study Program*, edited by F. O. SCHMITT AND F. G. WORDEN. Cambridge, Mass.: M.I.T. Press, 1974, pp. 205–214.

75. GRAYBIEL, A. M. Direct and indirect preoculomotor pathways of the brainstem: An autoradiographic study of the pontine reticular formation in the cat. *J. Comp. Neurol.*, 175: 37–78, 1977.

76. GROVES, P. M., MILLER, S. W., PARKER, M. V., AND REBEE, G. V. Organization by sensory modality in the reticular formation of the rat. *Brain Res.*, 54: 207–224, 1973.

77. HARDING, G. W., STOGSDILL, R. M., AND TOWE, A. L. Relative effects of pentobarbital and chloralose on the responsiveness of neurons in sensorimotor cerebral cortex of the domestic cat. *Neuroscience*, 4: 369–378, 1979.

78. HEATH, C. J., AND JONES, E. G. An experimental study of ascending connections from the posterior group of thalamic nuclei in the cat. *J. Comp. Neurol.*, 141: 397–426, 1971.

79. HEATH, C. J. AND JONES, E. G. The anatomical organization of the suprasylvian gyrus of the cat. *Ergeb. Anat. Entwicklungsgesch.*, 45: 1–64, 1971.

80. HEILMAN, K. M., PANDYA, D. N., AND GESCHWIND, N. Trimodal inattention following parietal lobe ablations. *Trans. Am. Neurol. Assoc.*, 95: 259–261, 1970.

81. HERNÁNDEZ-PÉON, R., AND HAGBARTH, K. E. Interaction between afferent and cortically induced reticular responses. *J. Neurophysiol.*, 18: 44–55, 1955.

82. HIND, J. E., ROSE, J. E., DAVIES, P. W., WOOLSEY, C. N., BENJAMIN, R. M., WELKER, W. I., AND THOMPSON, R. F. Unit activity in the auditory cortex. In: *Neural Mechanisms of the Auditory and Vestibular Systems*, edited by G. L. RASMUSSEN AND W. F. WINDLE. Springfield, IL: Thomas, 1960, pp. 201–210.

83. HOTTA, T., AND KAMEDA, K. Interactions between somatic and visual or auditory responses in the thalamus of the cat. *Exptl. Neurol.*, 8: 1–13, 1963.

84. HUANG, C. C., AND LINDSLEY, D. B. Polysensory responses and sensory interaction in pulvinar related postero-lateral thalamic nuclei in cat. *Electroenceph. Clin. Neurophysiol.*, 34: 265–280, 1973.

85. IMIG, T. J., AND REALE, R. A. Patterns of cortico-cortical connections related to tonotopic maps in cat auditory cortex. *J. Comp. Neurol.*, 192: 293–332, 1980.

86. IRVINE, D. R. F. Acoustic properties of neurons in posteromedial thalamus of cat. *J. Neurophysiol.*, 43: 395–408, 1980.

87. IRVINE, D. R. F., AND HUEBNER, H. Acoustic response characteristics of neurons in non-specific areas of cat cerebral cortex. *J. Neurophysiol.*, 42: 107–122, 1979.

88. ITOH, K., AND MIZUNO, N. Direct projections from the mesodiencephalic areas to the pericruciate cortex in the cat: an experimental study with the horseradish peroxidase method. *Brain Res.*, 116: 492–497, 1976.

89. ITOH, K., AND MIZUNO, N. Topographical arrangement of thalamocortical neurons in the centrolateral nucleus of the cat, with special references to a spino-thalamo-motor cortical path through the CL. *Exptl. Brain Res.* 30: 471–480, 1977.

90. JACKSON, G., AND IRVINE, D. R. F. Acoustic properties of neurons in cat mesencephalic reticular formation. *Proc. Aust. Physiol. Pharmacol. Soc.*, 9: 177P, 1978.

91. JASPER, H. H., AND AJMONE-MARSAN, C. *A Stereotaxic Atlas of the Diencephalon of the Cat.* Ottawa: National Research Council of Canada, 1954.

92. JONES, E. G. Some aspects of the organization of the thalamic reticular complex. *J. Comp. Neurol.*, 162: 285–308, 1975.

93. JONES, E. G., AND LEAVITT, R. Y. Retrograde axonal transport and the demonstration of non-specific projections to the cerebral cortex and striatum from thalamic intralaminar nuclei in the rat, cat and monkey. *J. Comp. Neurol.*, 154: 349–378, 1974.

94. JONES, E. G., AND POWELL, T. P. S. An analysis of the posterior group of thalamic nuclei on the basis of its afferent connections. *J. Comp. Neurol.*, 143: 185–216, 1971.

95. KAWAI, Y. Effect of destruction of the specific and nonspecific thalamic nuclei on the sensory responses in the cortical association area of the cat. *Psychiatria Neurol. Jap.*, 73: 584–615, 1971.

96. KAWAMURA, K. Corticocortical fiber connections of the cat cerebrum. I. The temporal region. *Brain Res.*, 51: 1–21, 1973.

97. KAWAMURA, K., BRODAL, A., AND HODDEVIK, G. The projection of the superior colliculus onto the reticular formation of the brain stem. An experimental study in the cat. *Exptl. Brain Res.* 19: 1–19, 1974.

98. KAWAMURA, K., AND HASHIKAWA, T. Cell bodies of origin of reticular projections from the superior colliculus in the cat: An experimental study with the use of horseradish peroxidase as a tracer. *J. Comp. Neurol.* 182: 1–16, 1978.

99. KENNEDY, H., AND BALEYDIER, C. Direct projections from thalamic intralaminar nuclei to extra-striate visual cortex in the cat traced with horseradish peroxidase. *Exptl. Brain Res.*, 28: 133–139, 1977.

100. KIANG, N. Y.-S. *Discharge Patterns of Single Fibers in the Cat's Auditory Nerve.* M.I.T. Monograph No. 35. Cambridge: M.I.T. Press, 1965.

101. KITSIKIS, A., AND STERIADE, M. Thalamic, callosal and reticular converging inputs to parietal association cortex in cat. *Brain Res.*, 93: 516–524, 1975.

102. KNIGHT, P. L. Representation of the cochlea within the anterior auditory field (AAF) of the cat. *Brain Res.*, 130: 447–467, 1977.

152   D. R. F. Irvine and D. P. Phillips

103. KNUDSEN, E. I., AND KONISHI, M. Space and frequency are represented separately in auditory midbrain of the owl. *J. Neurophysiol.*, 41: 870–884, 1978.

104. KREINDLER, A., CRIGHEL, E., AND MARINCHESCU, C. Integrative activity of the thalamic pulvinar-lateralis posterior complex and interrelations with the neocortex. *Exptl. Neurol.*, 22: 423–435, 1968.

105. LIU, Y.-M., AND SHEN, E. Pathways mediating the electrical response of the motor center to brief auditory and visual stimuli in cat. *Acta Physiol. Sin.*, 22: 104–118, 1958.

106. LOVE, J. A., AND SCOTT, J. W. Some response characteristics of cells of the magnocellular division of the medial geniculate body of the cat. *Can. J. Physiol. Pharmacol.*, 47: 881–888, 1969.

107. LYNCH, J. C., MOUNTCASTLE, V. B., TALBOT, W. H., AND YIN, T. C. T. Parietal lobe mechanisms for directed visual attention. *J. Neurophysiol.*, 40: 362–389, 1977.

108. MACCHI, G., BENTIVOGLIO, M., D'ATENA, C., ROSSINI, P., AND TEMPESTA, E. The cortical projections of the thalamic intralaminar nuclei restudied by means of the HRP retrograde axonal transport. *Neurosci. Lett.* 4: 121–126, 1977.

109. MACCHI, G., QUATTRINI, A., CHINZARI, P., MARCHESI, G., AND CAPOCCHI, G. Quantitative data on cell loss and cellullar atrophy of intralaminar nuclei following cortical and subcortical lesions. *Brain Res.*, 89: 43–59, 1975.

110. MANCIA, M., MECHELSE, K., AND MOLLICA, A. Microelectrode recording from midbrain reticular formation in the decerebrate cat.. *Arch. Ital. Biol.*, 95: 110–119, 1957.

111. MASSION, J., AND MEULDERS, M. Les potentiels évoqués visuels et auditifs de centre médian et leurs modification après décortication. *Arch. Intern. Physiol. Biochim.*, 69: 26–29, 1961.

112. MERZENICH, M. M., KNIGHT, P. L., AND ROTH, G. L. Representation of cochlea within primary auditory cortex in the cat. *J. Neurophysiol.*, 38: 231–249, 1975.

113. MERZENICH, M. M., AND REID, M. D. Representation of the cochlea within the inferior colliculus of the cat. *Brain Res.*, 77: 397–415, 1974.

114. MIZUNO, N., KONISHI, A., SATO, M., KAWAGUCHI, S., YAMAMOTO, T., KAWAMURA, S., AND YAMAWAKI, M. Thalamic afferents to the rostral portions of the middle suprasylvian gyrus in the cat. *Exptl. Neurol.*, 48: 79–87, 1975.

115. MOORE, R. Y., AND GOLDBERG, J. M. Ascending projections of the inferior colliculus in the cat. *J. Comp. Neurol.*, 121: 109–136, 1963.

116. MOUNTCASTLE, V. B., LYNCH, J. C., GEORGOPOLOUS, A., SAKATA, H., AND ACUNA, C. Posterior parietal association cortex of the monkey: command functions for operations within extrapersonal space. *J. Neurophysiol.*, 38: 871–908, 1975.

117. MURRAY, M. Degeneration of some intralaminar thalamic nuclei after cortical removals in the cat. *J. Comp. Neurol.*, 127: 341–367, 1966.

118. NAUTA, W., AND KUYPERS, H. Some ascending pathways in the brain stem reticular formation. In: *Reticular Formation of the Brain*, edited by H. JASPER, R. PROCTOR, R. KNIGHTON, W. NOSHAY, AND R. COSTELLO. London: Churchill, 1958, pp.3–30.

119. NELSON, C. N., AND BIGNALLL, K. E. Interactions of sensory and nonspecific thalamic inputs to cortical polysensory units in the squirrel monkey. *Exptl. Neurol.*, 40: 189–206, 1973.

120. NEWMAN, J. D., AND LINDSLEY, D. F. Single unit analysis of auditory processing in squirrel monkey frontal cortex. *Exptl. Brain. Res.* 25: 169–181, 1976.

121. O'BRIEN, J. H., AND FOX, S. S. Single cell activity in cat motor cortex. II Functional characteristics of the cell related to conditioning changes. *J. Neurophysiol.*, 32: 285–296, 1969.

122. O'BRIEN, J. H., AND ROSENBLUM, S. M. Influence of thalamic cooling on sensory responses in association cortex. *Brain Res. Bull.*, 4: 91–98, 1979.

123. OXBURY, J. M., CAMPBELL, D. C., AND OXBURY, S. M. Unilateral spatial neglect and impairments of spatial analysis and visual perception. *Brain*, 97: 551–564, 1974.

124. PALMER, L. A., ROSENQUIST, A. C., AND TUSA, R. J. The retinotopic organization of lateral suprasylvian visual areas in cat. *J. Comp. Neurol.*, 177: 237–256, 1978.

125. PAULA-BARBOSA, M. M., FEYO, P. B., AND SOUSA-PINTO, A. The association connexions of the suprasylvian fringe (SF) and other areas of the cat auditory cortex. *Exp. Brain Res.*, 23: 535–554, 1975.

126. PHILLIPS, D. P., AND IRVINE, D. R. F. Acoustic input to single neurons in pulvinar-posterior complex of cat thalamus. *J. Neurophysiol.*, 42: 123–136, 1979.

127. PHILLIPS, D. P., AND IRVINE, D. R. F. Responses of single neurons in physiologically-defined primary auditory cortex (AI) of the cat: Frequency tuning and responses to intensity. *J. Neurophysiol.* 45: 48–58, 1981.

128. PHILLIPS, D. S., DENNEY, D. D., ROBERTSON, R. T., HICKS, L. H., AND THOMPSON, R. F. Cortical projections of ascending nonspecific systems. *Physiol. Behav.*, 8: 269–277, 1972.

129. POGGIO, G. F., AND MOUNTCASTLE, V. B. A study of the functional contributions of the lemniscal and spinothalamic systems to somatic sensibility. *Johns Hopkins Med. J.*, 106: 266–316, 1960.

130. POWELL, T. P. S., AND COWAN, W. M. The interpretation of the degenerative changes in the intralaminar nuclei of the thalamus. *J. Neurol. Neurosurg. Psychiat.*, 30: 140–153, 1967.

131. POWELL, E. W., AND HATTON, J. B. Projections of the inferior colliculus in cat. *J. Comp. Neurol.*, 136: 183–192, 1969.

132. RASMINSKY, M., MAURO, A. J., AND ALBE-FESSARD, D. Projections of medial thalamic nuclei to putamen and cerebral frontal cortex in the cat, *Brain Res.*, 61: 69–77, 1973.

133. REALE, R. A., AND IMIG, T. J. Tonotopic organization in auditory cortex of the cat. *J. Comp. Neurol.*, 192: 265–292, 1980.

134. RINVIK, E. The corticothalamic projection from the gyrus proreus and the medial wall of the rostral hemisphere in the cat. An experimental study with silver impregnation methods. *Exptl. Brain Res.*, 5: 129–152, 1968.

135. RINVIK, E. Organization of thalamic connections from motor and somotosensory cortical areas in the cat. In: *Corticothalamic Projections and Sensorimotor Activities*, edited by T. FRIGYESI, E. RINVIK AND M. D. YAHR, New York: Raven, 1972, pp. 57–90.

136. RIOCH, D. McK. Studies on the diencephalon of carnnivora. I. The nuclear configuration of the thalamus, epithalamus, and hypothalamus of the dog and cat, *J. Comp. Neurol.*, 49: 1–119, 1929.

137. ROBERTSON, R. T. Thalamic projections to visually responsive regions of parietal cortex. *Brain Res. Bull.*, 1: 459–469, 1976.

138. ROBERTSON, R. T. Thalamic projections to areas 5 and 7 of parietal cortex in the cat. *Soc. Neurosci. Abstr.*, 4: 79, 1978.

139. ROBERTSON, R. T., MAYERS, K. S., TEYLER, T. J. BETTINGER, L. A., BIRCH, H., DAVIS, J. L., PHILLIPS, D. S., AND THOMPSON, R. F. Unit activity in posterior association cortex of cat. *J. Neurophysiol.*, 38: 780–793, 1975.

140. ROBERTSON, R. T., LYNCH, G. S., AND THOMPSON, R. F. Diencephalic distributions of ascending reticular systems. *Brain Res.*, 55: 309–322, 1973.

141. ROBERTSON, R. T., AND THOMPSON, R. F. Effects of subcortical ablations on cortical association responses in the cat. *Physiol. Behav.*, 10: 245–252, 1973.

142. ROBINSON, D. L., GOLDBERG, M. E., AND STANTON, G. B. Parietal association cortex in the primate: sensory mechanisms and behavioral modulations. *J. Neurophysiol.*, 41: 910–932, 1978.

143. ROSE, J. E., GREENWOOD, D. D., GOLDBERG, J. M., AND HIND, J. E. Some discharge characteristics of single neurons in the inferior colliculus of the cat. I. Tonotopical organization, relation of spike counts to tone intensity, and firing patterns of single elements. *J. Neurophysiol.*, 26: 294–320, 1963.

144. ROSE, J. E. GROSS, N. B., GEISLER, C. D., AND HIND, J. E. Some neural mechanisms in the inferior colliculus of the cat which may be relevant to the localization of a sound source. *J. Neurophysiol.*, 29: 288–314, 1966.

145. ROSE, J. E., AND WOOLSEY, C. N. Cortical connections and functional organization of the thalamic auditory system of the cat. In: *The Biological and Biochemical Bases of Behavior*, edited by H. F. HARLOW AND C. N. WOOLSEY. Madison: Univ. of Wisconsin Press, 1958, pp. 127–150.

146. RUTLEDGE, L. T., AND DUNCAN, J. A. Extracellular recording of converging input on cortical neurons using a flexible microelectrode. *Nature*, 210: 737–739, 1966.

147. SCHECHTER, P. B., AND MURPHY, E. H. Response characteristics of single cells in squirrel monkey frontal cortex. *Brain Res.*, 96: 66–70, 1975.

148. SCHEIBEL, M. E., AND SCHEIBEL, A. B. Patterns of organization in specific and nonspecific thalamic fields. In: *The Thalamus*, edited by D. P. PURPURA AND M. D. YAHR. New York: Columbia University Press, 1966, pp. 13–46.

149. SCHEIBEL, M. E., AND SCHEIBEL, A. B. The organization of the nucleus reticularis thalami: a Golgi study. *Brain Res.*, 1: 43–62, 1966.

150. SCHEIBEL, M. E., AND SCHEIBEL, A. B. Structural organization of nonspecific thalamic nuclei and their projection toward cortex. *Brain Res.*, 6: 60–94, 1967.

151. SCHEIBEL, M. E., AND SCHEIBEL, A. B. Input-output relations of the thalamic nonspecific system. *Brain, Behav. Evol.*, 6: 332–358, 1972.

152. SCHEIBEL, M. E., SCHEIBEL, A. B., MOLLICA, A., AND MORUZZI, G. Convergence and interaction of afferent impulses on single units of reticular formation. *J. Neurophysiol.*, 18: 309–331, 1955.

153. SHIMAZONO, Y., TORII, H., ENDO, M., IHARA, S., NARUKAWA, H., AND MATSUDA, M. Convergence of thalamic and sensory afferent impulses to single neurons in the cortical association area of cats. *Folia Psychiat. Neurol. Jap.*, 17: 144–155, 1963.

154. SLIMP, J. C., AND TOWE, A. L. Characteristics of somatic receptive fields of neurons in postcruciate cerebral cortex in awake-restrained and two anesthetic conditions in the same cat. *Soc. Neurosci. Abstr.*, 3: 492, 1977.

155. STEIN, B. E., AND ARIGBEDE, M. O. Unimodal and multimodal response properties of neurons in the cat's superior colliculus. *Exptl. Neurol.* 36: 179–196, 1972.

156. STERIADE, M., DIALLO, A., OAKSON, G., AND WHITEGUAY, B. Some synaptic inputs and ascending projections of lateralis posterior thalamic neurons. *Brain Res.*, 131: 39–53, 1977.

157. THOMPSON, R. F. Thalamocortical organization of association responses to auditory, tactile, and visual stimuli in the cat. *Internat. Congr. Physiol. Sci., Leiden*, 1962, p. 1057.

158. THOMPSON, R. F. Role of cortical association fields in auditory frequency discrimination. *J. Comp. Physiol. Psychol.*, 57: 335–339, 1964.

159. THOMPSON, R. F., JOHNSON, R. H., AND HOOPES, J. J. Organization of auditory, somatic sensory and visual projections to association fields of cerebral cortex in the cat. *J. Neurophysiol.*, 26: 343–364, 1963.

160. THOMPSON, R. F., AND SINDBERG, R. M. Auditory response fields in association and motor cortex of cat. *J. Neurophysiol.*, 23: 87–105, 1960.

161. THOMPSON, R. F., AND SMITH, H. E. Effects of association area lesions on auditory frequency discrimination in cat. *Psychon. Sci.*, 8: 123–124, 1967.

162. THOMPSON, R. F., SMITH, H. E., AND BLISS, D. Auditory, somatic sensory, and visual response interactions and interrelations in association and primary cortical fields of the cat. *J. Neurophysiol.*, 26: 365–378, 1963.

163. TOTIBADZE, N. K., AND MONIAVA, E. S. On the direct cortical connections of the nucleus centrum medianum thalami. *J. Comp. Neurol.*, 137: 347–360, 1969.

164. TSUCHITANI, C. Functional organization of lateral cell groups of cat superior olivary complex. *J. Neurophysiol.*, 40: 296–318, 1977.

165. VEDOVATO, M. Identification of afferent connections to cortical area 6aβ of the cat by means of retrograde horseradish peroxidase transport. *Neurosci Lett.*, 9: 303–310, 1978.

166. VERAART, C., MEULDERS, M., AND GODFRAIND, M.-M. Visual properties of neurons in pulvinar, nucleus lateralis posterior and nucleus suprageniculatus thalami in the cat. II. Quantitative investigation. *Brain Res.*, 44: 527–546, 1972.

167. WEBSTER, W. R. Central neural mechanisms of hearing. *Proc. Aust. Physiol. Pharmacol. Soc.*, 8: 1–7, 1977.

168. WEBSTER, W. R., AND AITKIN, L. M. Central auditory processing. In: *Handbook of Psychobiology*, edited by M. GAZZANIGA AND C. BLAKEMORE. New York: Academic, 1975, pp. 325–364.

169. WELCH, K., AND STUTEVILLE, P. Experimental production of unilateral neglect in monkeys. *Brain*, 81: 341–347, 1958.

170. WEPSIC, J. G. Multimodal sensory activation of cells in the magnocellular medial geniculate nucleus. *Exptl. Neurol.*, 15: 299–318, 1966.

171. WESTER, K. G., IRVINE, D. R. F., AND THOMPSON, R. F. Acoustic tuning of single cells in middle suprasylvian cortex of cat. *Brain Res.*, 76: 493–502, 1974.

172. WILSON, M. E., AND CRAGG, B. G. Projections from the lateral geniculate nucleus in the cat and monkey, *J. Anat., London*, 101: 677–692, 1967.

173. WISE, L. Z., AND IRVINE, D. R. F. Auditory response properties of neurones in intermediate and deep layers of cat superior colliculus. *Proc. Aust. Physiol. Pharmacol. Soc.*, 12: 18P, 1981.

174. WOOLSEY, C. N. Organization of cortical auditory system: a review and a synthesis. In: *Neural Mechanisms of the Auditory and Vestibular Systems*, edited by G. L. RASMUSSEN AND W. F. WINDLE. Springfield, IL: Thomas, 1960, p. 165–180.

Chapter 6

Functional Organization of the Auditory Cortex

Representation Beyond Tonotopy in the Bat

Nobuo Suga

Department of Biology, Washington University,
St. Louis, Missouri

1. Introduction*

1.1. Hypotheses and Neurophysiology of Cortical Representation of Auditory Information

The properties of an acoustic signal produced by an animal can be studied by examining the output of a microphone displayed on the screen of an oscilloscope. Such an examination, however, gives us

---

*The introduction, results and discussion of the DSCF area and the speculation about the function of specialized and unspecialized neurons are developed from Suga's reviews (32–34), while the discussion of the FM–FM and CF/CF areas is based upon our recent papers: Suga, O'Neill and Manabe ((41,42); O'Neill and

only limited information about the signal properties, so that the signal is usually analyzed with a spectrum analyzer. The spectrum analyzer has many filters tuned to different frequencies and it expresses the output of each as a function of time. Therefore, the properties of the acoustic signals are expressed by a pattern that appears in three coordinates: frequency, amplitude and time. To recognize individual acoustic patterns with an instrument, information-bearing elements (elements characterizing the signal) are first extracted and then their combinations are examined. The question next arises: how are acoustic signals analyzed and processed in the auditory system?

In the mammalian cochlea, sensory cells may be viewed as filters arranged along the basilar membrane for frequency analysis and their outputs are coded by primary auditory neurons that discharge action potentials at higher rates for larger outputs. Therefore, at the periphery, the frequency of an acoustic signal is expressed by the location of activated neurons and its amplitude by their discharge rates. In order to clarify the significance of recent discoveries, it must be stressed that the peripheral auditory system has an anatomical axis only for frequency and that activity of individual neurons cannot uniquely express the properties of an acoustic signal. For instance, a peripheral neuron tuned to 40 kHz responds not only to a pure tone of 40 kHz, but also to an FM sound sweeping across 40 kHz regardless of sweep direction and to a noise burst containing 40 kHz regardless of bandwidth. Therefore, the neuron cannot code the type of acoustic signal stimulating the ear. *The properties of an acoustic signal are appropriately expressed only by the spatiotemporal pattern of activity of all peripheral neurons.*

Action potentials sent into the brain by peripheral neurons are transmitted to many auditory nuclei and finally to the cerebral cortex, where a multiple representation of the cochlea is prominent. This multiple represenatation suggests that separate auditory areas are concerned with representation of different types of auditory information. In this article, I shall stress the functional organization of the auditory cortex of the mustached bat, *Pteronotus parnellii rubiginosus*, rather than the multiple representation per se, because in the areas where different parts of the cochlea project with overlap, neurons are sensitive to particular combinations of elements in complex sounds. Therefore the functional organization within these areas is undoubtedly much more important than its tonotopic representation for understanding the neural mecha-

---

Suga *(19)* and Suga and O'Neill *(39)*. The IBP filter hypothesis is a translation of part of Suga's review *(34)*. References cited are minimized to those that are most related to this article.

nisms for processing complex sound. To examine the functional organization, we should first consider hypotheses on how the cerebral auditory cortex is functionally organized to represent auditory information by neural activity and what response properties the individual neurons, related directly to recognition of auditory information, exhibit. Concerning the neural basis of recognition of acoustic signals, at least four working hypotheses are conceivable. They are probably not exclusive, but individually valid depending upon the types of auditory information and species.

The amplitude spectrum hypothesis suggests that recognition of an acoustic signal is directly related to the spatiotemporal pattern of activity of "nonspecialized" neurons arranged along the coordinates of frequency vs amplitude at a hypothetical auditory center and is not related to the activity of neurons which respond selectively to the signal itself (Fig. 6.1, part 1). According to this hypothesis, each neuron has a small excitatory area* tuned to a particular frequency and amplitude and responds to many different types of sounds containing a component that stimulates this area.

The representation of acoustic signals in the cerebral cortex was first studied in the cat by Woolsey and Walzl (49). Since then, a frequency axis, which they described, has been demonstrated in many different species of mammals, including the mustached bat, *Pteronotus parnellii rubiginosus* (36), the cat (51) and monkeys (50). A threshold or amplitude axis, however, has not yet been reported in any animal, with the exceptions of the dog (48) and the mustached bat (31).

Threshold representation by neurons with a monotonic impulse-count function is different from *amplitude representation* by neurons with a nonmonotonic impulse-count function. The former has three disadvantages compared to the latter. In threshold representation, (i) more neurons are involved in representing the amplitude spectrum of a stronger acoustic stimulus, (ii) the boundary between excited and nonexcited neurons is unclear and fluctuates because neural responses usually fluctuate near and at the threshold, and (iii) no lateral inhibition mechanism operates to improve the contrast of the boundary between excited and nonexcited neurons.

The threshold representation of sound amplitude found in the dog by Tunturi (48) has always been intriguing, but it has not been accepted by auditory physiologists, probably for the following four reasons: (i) the data were obtained from deeply anesthetized ani-

---

*The area above a frequency-tuning curve is called a response area. Responses of auditory neurons in the central nervous system to stimuli can be excitatory or inhibitory, so that there are excitatory and inhibitory response areas. These are called the *excitatory* and *inhibitory areas*, respectively.

Representation of acoustic signals in a hypothetical auditory center

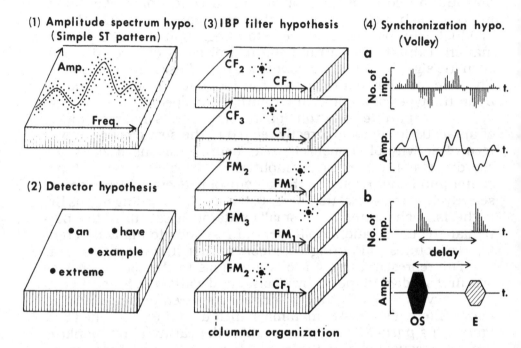

FIG. 6.1. Four working hypotheses for representation of auditory information by neural activity in a hypothetical center for recognition of acoustic signals. (1) The *amplitude-spectrum hypothesis*, which may also be called the simple spatiotemporal pattern hypothesis. (2) The *detector hypothesis*. (3) *The IBP filter hypothesis*; here each coordinate is formed by neurons tuned to certain combinations of information-bearing parameters (IBPs) that characterize biologically important signal elements. In all these three hypotheses, the properties of acoustic signals are represented by the spatiotemporal pattern of neural activity. The properties of individual neurons and the interpretation of their functions are, however, different according to the hypotheses. (4) The *synchronization hypothesis*, which is the hyperinterpretation of the Wever's volley theory. In 4a, the lower trace represents sound wave and the upper trace represents the compound period histogram of a single neuron response to it. In 4b, the lower trace represents an orientation sound (OS) and an echo (E), and the upper trace, the PST histogram of a single neuron response (see text).

mals after the application of strychnine sulfate to the auditory cortex; (ii) no one has yet demonstrated threshold representation in the auditory cortex of other species; (iii) the threshold representation was found only for the stimulation of the ipsilateral ear, which does not supply the main excitatory input to the auditory cortex. The main excitatory input is contralateral and is uniformly lower in threshold than that from the ipsilateral ear; and (iv) the functional significance of the threshold representation is not clear, because most of the cortical auditory neurons show relatively phasic on-responses to tone bursts and nonmonotonic impulse-count functions. Most DSCF neurons in the mustached bat show a nonmonotonic impulse-count function and are tuned to particular stimulus amplitudes. Comparable data have been apparently obtained in the primary auditory cortex of the macaque monkey and vague suggestions of amplitopic representation were made (50). But it has not yet been demonstrated in animals other than the mustached bat (31). As summarized below, the coordinates of frequency vs amplitude have thus far been demonstrated only in the DSCF area of the mustached bat (Fig. 6.8).

The detector hypothesis suggests that recognition of a biologically important acoustic signal is directly related to the excitation of detector neurons that selectively respond to that particular signal (Fig. 6.1, part 2). Different types of detector neurons are, of course, arranged in a particular spatial pattern in the hypothetical center. The spatiotemporal pattern of neural activity in the center will thus change according to the sequence of biologically significant sounds. The essential distinction of the detector hypothesis is a one-to-one correspondence between a categorical perception and a detector neuron or a group of detector neurons. Neurons specialized to respond selectively to certain types of acoustic stimuli have been found in a few different species of mammals (in bats, 19, 30, 39–42). It is a future problem to ascertain whether some "specialized" neurons that have been found thus far can be appropriately called detectors or not. The response properties of specialized neurons recently discovered in the auditory cortex of the mustached bat and the functional organization of the cortex fit the third hypothesis, which is described below.

The IBP filter hypothesis falls between the two hypotheses described above. In auditory neurophysiology, the statement that neurons respond to "sound X" but not to others, is not quantitative, so that we usually study the filter properties of the neurons by measuring tuning curves for sound X by changing its individual parameters (19, 31, 39–42). Therefore, we can treat the neurons as filters. This is a theoretical advantage, since a filter acts as a kind of cross-correlator. Neurons are filters that correlate acoustic signals with their filter properties, i.e., stored information, and the degree of correlation is expressed by the magnitude of the output of the filters. In other words, neurons are maximally excited only when the properties of acoustic signals perfectly match their filter properties. All neurons in the auditory system, including peripheral ones, act as filters. *Specialized neurons expressing the outputs of neural circuits tuned to particular information-bearing parameters*

(IBPs) (32) or *particular combination of IBPs are called IBP filters. An IBP is the limited part of a continuum that carries information important for the species.**

The IBP filter hypothesis states that the recognition of auditory information is directly related to the spatiotemporal patterns of activities of many specialized neurons, i.e., the spatiotemporal pattern of the outputs of many IBP filters that are systematically arranged in a hypothetical center and that act as a kind of cross-correlator (Fig. 6.1, part 3; 34). The data obtained from the auditory cortex of the mustached bat (19, 31, 39–42) complement this hypothesis and add the following four statements, which are important in understanding neural processing of complex acoustic signals: (i) complex sound is processed by neurons specialized for examining, i.e., IBP filters tuned to different combinations of signal elements; (ii) different types of IBP filters are aggregated separately in identifiable areas of the cerebral cortex; (iii) in each aggregate, IBP filters are arranged along axes for the systematic representation of IBPs, i.e., signal variation, which has biological importance; and (iv) the axis—population of neurons—representing an IBP is apportioned according to the biological importance of the IBP (Fig. 6.1, part 3; 53, 54).

After the presentation of the data obtained from the mustached bat, I shall discuss the IBP filter hypothesis in detail.

The synchronization hypothesis is the hyperinterpretation of Wever's volley theory (55). When information-bearing elements are lower than 5 kHz, peripheral neurons produce discharges synchronized with the sound waves. The envelope of a compound period histogram of a single unit response thus reproduces the stimulus wave form (Fig. 6.1, part 4a). The synchronization hypothesis states that neurons in the hypothetical auditory center respond in the same manner as peripheral neurons and that synchronous discharges themselves are directly related to the recognition of acoustic signals. In small mammals, many of the predominant components in their communication sounds are higher than 5 kHz, so that synchronous discharges may play only a limited role in signal recognition. Echo-locating bats emit biosonar signals, which are usually higher than 10 kHz, and listen to echoes. The delay of an echo from the emitted sound is the primary cue for target ranging. Peripheral neurons show discharges synchronized with

---

*IBPs include not only parameters characterizing information-bearing elements of a complex sound, but also interaural time and amplitude differences, interval between signals (e.g., echo delay) and other parameters characterizing combinations of information-bearing elements that are important for communication and/or echolocation. An identical IBP can be quite different in its biological significance for different species of animals.

each emitted sound and echo. Therefore, range information is coded by the interval between a pair of grouped discharges (Fig. 6.1, part 4b). The synchronization hypothesis further states that range perception is a direct consequence of paired group discharges, not that resulting from an excitation of neurons specialized for responding to particular echo delays.

Neurons in the auditory cortex of mammals show no discharges synchronized with sound waves higher than 1 kHz. In the FM-FM area of the auditory cortex of the mustached bat, a time interval between acoustic stimuli is represented by a locus or distribution of activated neurons (39). Therefore, the synchronization hypothesis has a limitation to its validity and can be true only for recognition of a part of the properties of acoustic signals. Synchronous (phase-locked) discharges play an important role in processing information for sound localization (e.g., 52). There is, however, a possibility that the location of a sound source is not represented by a magnitude of phase-locked discharges itself, but by a locus or distribution of activated neurons that are tuned to particular interaural time (phase) differences.

The above four hypotheses can be tested by examining the functional organization of the cerebrum for the representation of auditory information. The functional organization may be different among species that use different types of acoustic signals and extract auditory information for different purposes. The same or similar functional organization found among different species may have different biological significance for individual species. In this article, I will first summarize the data obtained from the mustached bat and then will discuss the theories in relation to the data.

1.2. Unique Aspects of the Biosonar Signals and the Peripheral Auditory System of the Mustached Bat

Since the auditory system has evolved for processing acoustic signals that are important for the species, the functional organization of the auditory system can be most effectively explored when research is performed with regard to these signals. Accordingly, I will first summarize the properties of the biosonar signals of the mustached bat.

For prey capture and short-range navigation, the mustached bat produces *biosonar signals (orientation sounds)*, each consisting of a long constant-frequency (CF) component followed by a short frequency-modulated (FM) component (Fig. 6.2A). Each component is composed of up to four harmonics (H_1–H_4) and, therefore, there can be a total of up to eight components (CF_{1-4},

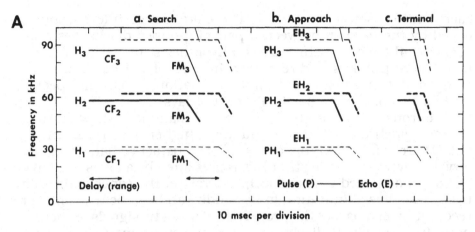

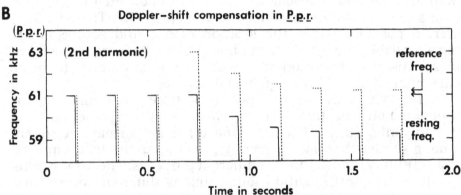

FIG. 6.2. A. Schematized sonagrams of the synthesized mustached bat's orientation sounds (P; solid lines) and echoes (E; dashed lines) mimicking those in the three phases of target-directed flight. The three harmonics of both the orientation sounds (PH$_{1-3}$) and the echoes (EH$_{1-3}$) each contain a long CF component (CF$_{1-3}$) followed by a short FM component (FM$_{1-3}$). The fourth harmonic (H$_4$) is not shown in the figure. (a) Search phase: CF and FM durations are 30 and 4 ms, respectively. (b) Approach phase: CF and FM durations are 15 and 3 ms. (c) Terminal phase: CF and FM durations are 5 and 2 ms. Repetition rates for a, b and c are 10, 40 and 100 pairs/s, respectively. The thickness of the lines indicates the relative amplitudes of each harmonic in the orientation sounds (pulses) and echoes: H$_2$ is strongest, followed by H$_3$ (−6 to −12 dB) and H$_1$ (−12 to −24 dB). Echo delay is measured as the time interval between the onsets of corresponding components of the orientation sound and the echo in a stimulus pair.

FIG. 6.2. B, Sonagrams for explanation of Doppler-shift compensation in the mustached bat. For simplicity, only the second harmonics are shown. An echo (dashed line) returns with a short delay after each emitted orientation sound (solid line). The echoes for the first three emitted sounds are not Doppler-shifted, while those for the rest are Doppler-shifted by 2.0 kHz. Note the difference between the resting and reference

FM_{1-4}). While in flight, the bat produces orientation sounds at various repetition rates and durations. When the bat is hunting but has not detected a target *(search phase)*, the sounds are 20–30 ms in duration and are repeated 5–10 times per second. When the bat detects and approaches a target, the sound duration often initially increases by about 10 ms and then always decreases roughly linearly with distance. During this *approach phase*, the repetition rate increases to 30–40/s. Just before intercepting the target *(terminal phase)*, the repetition rate increases dramatically to 90–100/s and the sounds shorten to 5–7 ms (18,20; Suga, et al., unpublished). During the search phase, the frequency of the predominant second harmonic (H_2) remains at about 61 kHz for 18–26 ms, then sweeps down to about 50 kHz within 2–4 ms. Since the reflected sound energy is highly concentrated at a single wavelength, the long CF tone is ideal for target detection and measurement of target velocity. The FM component, on the other hand, is more appropriate for ranging, localizing and characterizing a target, because of its short duration and wide frequency bandwidth. In orientation sounds recorded in our laboratory, the first harmonic (H_1) is always much weaker than H_2 and H_3.

The mustached bat, like the horseshoe bat, *Rhinolophus ferrumequinum*, performs a fascinating acoustic behavior called *Doppler-shift compensation* (20). When there are no objects moving relative to the bat, it produces sounds with the second harmonic CF (CF_2) at about 61 kHz. If the bat receives a Doppler-shifted echo, say at 63 kHz, from an approaching object, it reduces the frequency of the subsequent orientation sounds so as to stabilize the echo CF_2 at several hundred Hz above 61 kHz (Fig. 6.2B). Doppler-shift compensation takes place for both stationary (e.g., wall) and moving (e.g., moving hand) objects. Behaviorally, the bat is extraordinarily sensitive to echoes from fluttering targets such as flying insects (7).

To analyze the small shifts in the frequency of echo CF_2 produced by relative motion and flutter of a target, the cochlea of the mustached bat is apparently specialized for fine frequency analysis of the CF_2 in orientation sounds and echoes. The cochlear microphonic response is sharply tuned to 61 kHz (Fig. 6.3B) and shows prominent ringing at frequencies at or near 61 kHz (Fig. 6.3A). As a consequence, single auditory nerve fibers most sensitive to 60–63

FIG. 6.2. (continued)

frequencies. The resting frequency is the frequency of an emitted CF signal, when there are no Doppler-shifted echoes. The reference frequency is that of an echo CF at the steady state of Doppler-shift compensation. [Based on Schnitzler (20) and personal observations.]

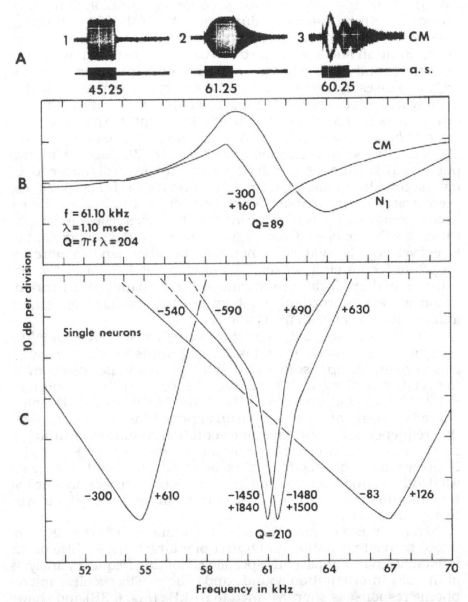

FIG. 6.3. A, Cochlear microphonic (CM, upper traces) evoked by 2.0 ms tone bursts (a.s., lower traces). The frequencies of the tone bursts are indicated by the figures below them in kHz. In A2, the CM shows prominent ringing, which is called CM-aft. In A3, the CM-on shows 1.0 kHz beat, which corresponds to the difference between the stimulus and resonance frequencies (Suga and Jen, 1977, ref. 37). B, Frequency-tuning curves of CM and N_1 recorded from the rim of the round window. The slopes and Q of the CM threshold curve around 61 kHz, the resonance frequency (f) and time constant (λ) of CM-aft, and Q in terms of CM-aft are given within the

kHz, especially 61 kHz, show extremely sharp frequency-tuning (Fig. 6.3C). The slopes on each side of the tuning curve between the minimum threshold and 30 dB above it are -1450 and $+1840$ dB/octave. The slopes between 30 and 60 dB above the minimum threshold are less, however, -540 and $+690$ dB/octave. This means that the extremely sharp tuning curves are the result of two filters. The filter responsible for the lower part of these tuning curves has a Q-value of 210 on the average and probably results from the sharply tuned local resonator that is responsible for the ringing in the cochlear microphonic response. The filter responsible for the upper part of the tuning curves has a Q-value of 70 on the average and presumably results from resonators that are strongly coupled with each other in the longitudinal direction of the basilar membrane and that are activated by the traveling wave (37). It is not yet known what mechanism is responsible for the sharply tuned local resonator in the cochlea. To a lesser extent, frequency-tuning curves of neurons sensitive to the CF_1 and CF_3 are also sharper than others tuned to different frequencies (37, 47). Neurons sharply tuned to sounds between 60 and 63 kHz apparently function in fine frequency analysis of CF_2. When the target is a flying insect, the returning echo CF can be modulated in frequency and amplitude by the beating wings. *The 60–63 kHz sensitive neurons can code frequency modulation as low as ±0.01%, i.e., ±6.1 Hz shift for 61 kHz,* by changing their discharge rate synchronously with the modulation (37). These neurons can also code a minor amplitude modulation occuring at a high rate of up to 3.0 kHz (47) (see ref. 32 for a detailed summary of the peripheral neurons).

At the periphery, *all neurons act as a filter in the frequency domain* and are systematically arranged along a certain frequency axis, which has an anatomical correlate. *Except for frequency, however, all other acoustic parameters, amplitude, echo delay, etc., have no anatomical correlate and are expressed by temporal patterns of impulse discharges.* In the auditory cortex, on the other hand, not only frequency, but also other acoustic parameters carrying information important for the animal, are systematically represented by the location or distribution of activated neurons.

FIG. 6.3. (continued)
graph. Q (quality factor) is the best frequency divided by the bandwidth of the threshold curve at 3 dB above the minimum threshold. C, Frequency-tuning curves of single neurons at the periphery. Their slopes are given in dB per octave. The Q of the tuning curves at about 61 kHz is 210. Each of these curves is the average of curves obtained from several mustached bats lightly anesthetized with sodium pentobarbital (Suga, 1978, ref. 32).

2. Methods*

2.1. Materials

Mustached bats, *Pteronotus parnellii rubiginosus†* (family Mormoopidae) from Panama were used. Their body weight is about 22 g.

2.2. Surgery

For surgery, the bats were injected with either chlorpromazine followed by sodium pentobarbital (30 mg/kg body weight) or Fentanyl–Droperidol mixture (Innovar-Vet 0.08 mg/kg body weight). The dorsal part of the skull was exposed and the flat head of a 1.5 cm-long nail was mounted onto it with glue and cement.

2.3. Animals During Experiments

Each bat was put in a lucite restraint that was suspended with a rubber band at the center of a sound-proofed room. To immobilize the bat's head, the nail was locked into a metal rod with a set screw. Most experiments on the DSCF area and some experiments on the FM–FM and CF/CF areas were performed with lightly anesthetized animals, whereas some experiments on the DSCF area and most experiments on the FM–FM and CF/CF areas were performed with unanesthetized animals. By "unanesthetized" animals, we mean those who did not receive any general anesthetic for at least 48 h before and during the experiments and who could eat, drink and fly voluntarily. If the animals were anesthetized, the sound-proofed room was kept at about 35°C. If they were unanesthetized, the room was maintained at 25–28°C. The unanesthetized animals had water and sometimes crushed mealworms during the experiments and their surgical wound was treated with local anesthetic (Xylocaine). After the 1-day experiments, the animals were returned to the animal room. The same individuals were used for up to 8 weeks (16 1-day sessions).

2.4. Acoustic Stimuli

Constant frequency (CF) tones, frequency-modulated (FM) sounds and noise bursts (NB) were delivered singly or in combinations

---

*The essential parts of the methods are briefly described. See the original papers cited for details.

†*P. parnellii rubiginosus* was previously called *Chilonycteris rubiginosa*. Mustached bats from different Central American populations may have slightly different resting frequencies in their biosonar signals.

from a condenser loudspeaker placed at 72 cm in front of the bat (see ref. 30 for details). All sound amplitudes in this article are expressed in dB SPL (sound pressure level in decibels referred to 0.0002 dyne/cm² rms). The duration, rise–decay time and repetition rate of stimuli were 4–50 ms, 0.5 ms and, 1.5–100/s, respectively. To reduce echoes, the inner wall of the sound-proofed room was covered with convoluted foam rubber and the micromanipulators and metal rods around the bat's head were covered with convoluted foam or cheese cloth.

2.5. Data Acquisition

At the initial stage of our research on tonotopic and amplitopic representation, a large tungsten wire electrode (10–30 μm tip) was mainly used for quick survey of the auditory cortex. Responses of a large cluster of neurons were full-wave rectified and then averaged with a computer, while responses of a single neuron in a small cluster were plotted as peristimulus–time (PST) histograms with a computer after selection of action potentials of uniform amplitudes with an amplitude discriminator (31, 36). Except for this initial stage, the research was performed with a small tungsten wire electrode (5–15 μm tip). Although the signal-to-noise ratio was high (more than 10 dB), amplitude discrimination was always used. Responses of a single neuron were expressed by PST and/or cumulative histograms.

2.6. Anesthetics and Neural Activity

The effect of an anesthetic would be prominent on neural activity in the auditory cortex and there is no doubt that neurophysiological experiments can be performed to clearly demonstrate the effect of an anesthetic. It is, however, obvious that its effect depends upon its dose and decreases with time after the administration, if it is intermittent. When the bats were anesthetized for surgery, recording of single unit activity was started at least 3 h after the initial injection of 30 mg sodium pentobarbital/kg body weight. During the recording, the bat usually showed various reflexes, voluntary movements and occasional sound emission. When the bat started to move frequently, an additional dose of 10 mg sodium pentobarbital/kg body weight was given and local anesthetic was applied to the surgical wound. Responses of a cluster of a few neurons to acoustic stimuli were comparable before and after such updates. If an electrode was placed at nearly the same site in the auditory cortex of the same animal a few days or a few weeks later, using no general anesthesia, the data obtained were comparable to those obtained from the lightly anesthetized bat. In our experiments with lightly anesthetized animals, therefore, the effect of sodium pentobarbital was apparently minimized. Since we could not tell the difference between the data obtained from the lightly anesthetized bats and unanesthetized, awake bats, we pooled the data from both. Individual data are, however, marked with "UA" (unanesthetized) or "A" (anesthetized). The numbers following "A" indicate the number of hours after the administration of sodium pentobarbital that the data were obtained (e.g., Figs. 6.11 and 6.15).

3. Results and Discussion

3.1. Areas Specialized for Processing Biosonar Information in the Cerebrum

The cerebral auditory cortex of the mustached bat is about 0.9 mm thick and, as in other mammals, consists of six layers. Therefore, it is expected that there are afferent and efferent fibers, as in the auditory cortex of cats and monkeys (Fig. 6.4B). The density of neurons in the cortex appears to be nearly the same in the bat and the cat (E. G. Jones, unpublished). The auditory cortex of the mustached bat is very large and contains at least three major districts specialized for processing different types of biosonar information: DSCF, FM–FM and CF/CF areas (Fig. 6.4A). The *DSCF (Doppler-shifted-CF processing) area* reflects the remarkable specialization of the peripheral auditory system and has two axes representing either echo frequency (target velocity information) or amplitude (subtended target angle information). The *FM–FM area* consists of neurons examining combinations of FM components in orientation sounds and echoes and has an axis representing target range. The *CF/CF area*, on the other hand, is comprised of neurons examining combinations of harmonically or quasi-harmonically related CF components in orientation sounds and echoes. The cerebral cortex of the mustached bat shows fascinating functional organization for processing biosonar information. In the following, I will describe how information about target velocity, subtended angle, and range are represented by neural activity in the cerebrum.*

---

*Since a barbiturate can have a drastic effect on neural activity in the cerebrum and an overdose can completely stop it, doubt persists about the extent to which the neural activity observed in our lightly anesthetized animals is normal. Some argue that the systematic tonotopic and amplitopic representation found in lightly anesthetized animals (31,36) can be demonstrated only with the anesthetized animals and that such representation can be demonstrated only with activity

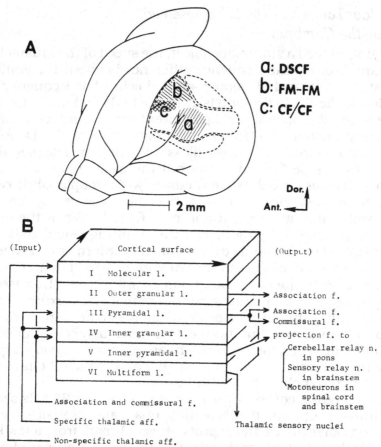

FIG. 6.4. A, Dorsolateral view of the cerebrum of the mustached bat. The areas within the dashed lines are the auditory cortex. Three areas specialized for processing biosonar signals have thus far been found. These are the DSCF, FM–FM and CF/CF areas indicated by a, b and c, respectively. B, The bat's auditory cortex consists of six layers and probably receives afferent fibers (arrows on the left) and sends out efferent fibers (arrows on the right), as that of other mammals.

recorded from the fourth cortical layer. I find these arguments inappropriate, because the systematic changes in best frequency and best amplitude along the cortical surface were demonstrated in oblique electrode penetrations with unanesthetized, awake mustached bats, and also because columnar organization in terms of best frequency and best amplitude was demonstrated in orthogonal electrode penetrations (38). As reviewed in this article, the data on FM–FM and CF/CF areas obtained from the unanesthetized bats also fit those obtained from the lightly anesthetized bats. Contrary to my expectations, cortical neurons specialized to respond to particular combinations of signal elements are as active in the lightly anesthetized animals as in those that are unanesthetized and awake.

3.1.1. How Target Velocity is Represented
by the Cerebrum

When flying toward a stationary object, the speed of the mustached bat is expressed by the frequency difference between the emitted sound and the returning Doppler-shifted echo. The frequency information of the emitted sound is available to the bat in the form of vocal self-stimulation and perhaps efferent copy, about which little is currently known. The frequency information of the Doppler-shifted echo is available regardless of whether Doppler-shift compensation is performed or not but the measurement of echo frequency becomes much more accurate with Doppler-shift compensation, because the echo frequency is then analyzed by the very sharply tuned neural filters at and near 61 kHz. When the mustached bat pursues a flying insect, the resulting Doppler-shifted echo consists of both dc and ac components; the former is related to the velocity difference between the bat and the insect, whereas the latter (periodic frequency modulation) results from the insect's beating wings. The frequency of the dc component is expressed by the spatial and temporal distribution of activated neurons with different best frequencies, but its analysis is obviously confounded by the ac component. The detection of the ac component is very important for the bat and it is coded by spatial and temporal change in neural activity synchronized with flutter.

In the cerebral cortex of the mustached bat, 61–63 kHz sensitive neurons occupy about 30% of the primary auditory cortex, even though the range of hearing is probably very broad, from a few kHz to 150 kHz (Fig. 6.5). Since this 61–63 kHz tuned area is undoubtedly the area that is devoted to processing information carried by the Doppler-shifted-CF component of echoes, it is named the *DSCF area*. In the DSCF area, each orthogonal microelectrode penetration is characterized by neurons with nearly identical *best frequencies (BFs)* and excitatory areas (excitatory-frequency-tuning curves). The best frequency characterizing individual cortical columns systematically varies with the location on the cortical surface, so that there is systematic *tonotopic representation:* 61 kHz-sensitive neurons are located at the center and 63 kHz-sensitive neurons at the circumference (Figs. 6.5 and 6.8; 36).* Neurons with best frequencies of 64–65 kHz are found along the ventral

---

*The tonotopic representation in the DSCF area correlates with the "resting frequency" of the CF_2 emitted by the individual mustached bat. When a bat's resting frequency is found to be about 60.5 kHz instead of 61.0 kHz, its DSCF area represents sounds mainly between 60.5 and 62.5 kHz. On the other hand, if the resting frequency of another bat is about 61.5 kHz, its DCSF area represents sounds mainly between 61.5 and 63.5 kHz. Thus, differences between individual bats in the frequency of the biosonar signal are reflected by the DSCF area (40).

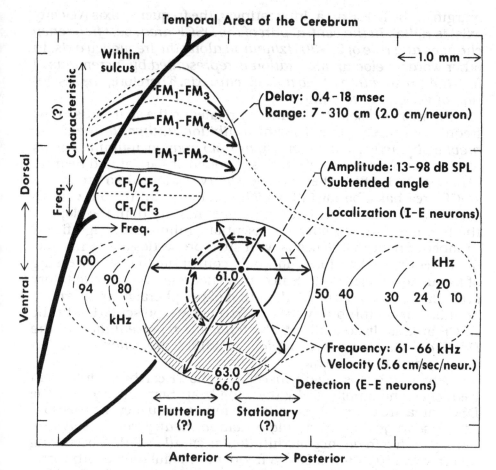

FIG. 6.5. Graphic summary of the functional organization of the auditory cortex. The tonotopic representation of the primary auditory cortex and the functional organization of the DSCF, FM–FM and CF/CF areas are indicated by lines and arrows. The DSCF area has axes representing either target velocity or subtended angle information and is divided into two subdivisions suited for either target detection or localization. Its anterior and posterior halves may be suited for processing echoes from either fluttering or stationary targets. The FM–FM area consists of three major types of FM_1–FM_n facilitation neurons and has an axis representing target range, orthogonal to which target characteristics may be represented. The CF/CF area consists of two major types of CF_1/CF_n facilitation neurons that are sensitive to combinations of harmonically or quasiharmonically related CF components and has the frequency vs frequency coordinates. The CF/CF area probably represents velocity information systematically. Note that there are nonprimary auditory cortices dorsal and also ventral to the primary auditory cortex shown in the figure (Fig. 6.4A) and that the functional organization of these nonprimary auditory cortices has not yet been studied. The broad lines are arteries, one of which (long) is on a sulcus. See the text for detail.

margin of the DSCF area. Interestingly, the frequency axis (velocity axis) is radial. *In the ventral part of the DSCF area, best frequency changes at a rate of 20–30 Hz/neuron along the frequency axis.* In other words, *velocity information is represented by increments of 5.6–8.4 cm/s/neuron along the velocity axis.* Therefore, the resolution of velocity is very fine.

Frequency resolution is directly related to the sharpness of the frequency-tuning curve. *Lateral inhibition further sharpens the previously mentioned extremely narrow frequency-tuning curves of peripheral neurons,* whose best frequencies are at and near 61 kHz (Fig. 6.6; 38).† The narrowest curve obtained so far in the DSCF area has a bandwidth which remains within 300 Hz over a broad amplitude range. Because of lateral inhibition, the slope of the frequency-tuning curve is essentially infinite in a significant number of neurons. Neurons with such properties act as narrow-band frequency detectors and would contribute to fine processing of frequency information. Since many neurons have an excitatory area that is much narrower than that of peripheral neurons and is bounded by an inhibitory area(s), frequency representation in the DSCF area is discrete (31).* *The tonotopic representation in the DSCF area is thus not a simple unmodified projection of the frequency axis in the cochlea.*

When a target is a flying insect, its echo CF can be modulated in frequency and amplitude by the wing beat. Some neurons in the DSCF area are extremely sensitive to minor frequency modulation. *The discharge rate of a single neuron clearly varies synchronously with a frequency modulation as small as 0.01%* occurring at a rate of 100/s (38). A 0.01% frequency-modulation depth corresponds to a 6.1 Hz frequency shift at 61 kHz. Since the Doppler-shift by the wing beat of a moth can be as large as 800 Hz, it is quite possible that the wing beat information is processed by synchronous discharges of such neurons and also by synchronous change in excitation from one group of neurons to another with different best frequencies and best amplitudes in the frequency vs amplitude coordinates, which will be described in the following section. The DSCF area also contains neurons insensitive to the ac component of Doppler-shift. Their excitatory areas are very narrow and are probably sandwiched in between inhibitory areas. These neurons may be considered to be specialized for detection of the dc compo-

---

†It should be noted that a neural mechanism for sharpening of a frequency-tuning curve operates even in the cochlear nucleus (47) and that the mechanism was first demonstrated in cats by Katsuki et al. (13).

*The same neural mechanism for sharpening of the cortical representation of a sensory signal was first hypothesized in the somatosensory system by Mountcastle (16).

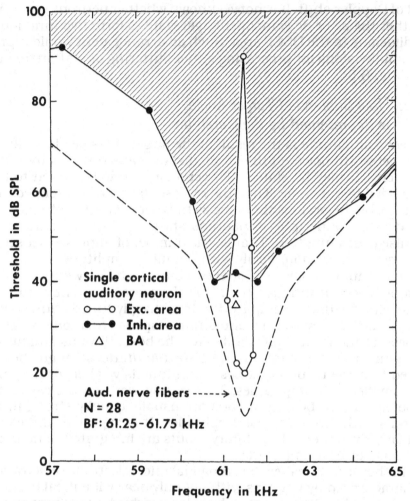

FIG. 6.6. The excitatory (open circles, unshaded) and inhibitory (solid circles, shaded) areas of a single neuron in the DSCF area. Since the rate of spontaneous discharge was not high enough to measure the inhibitory area with a single tone stimulus, a conditioning tone burst was delivered prior to an excitatory tone burst of 61.3 kHz and 37 dB SPL (cross). The conditioning sound was changed in frequency and amplitude to measure the inhibitory area, in which excitation evoked by the 61.3 kHz tone burst was inhibited by the conditioning sound. The inhibitory area is on both sides of the excitatory area and greatly overlaps it. The neuron showed phasic on-response followed by inhibition to tone bursts in the excitatory area overlapping the inhibitory area. The best amplitude of the neuron was 33 dB SPL (triangle). For comparison, the averaged excitatory area of 28 single cochlear nerve fibers with best frequencies between 61.25 and 61.75 kHz is shown (dashed line). Note that the excitatory area of the cortical neuron is much sharper than that of the peripheral neurons and that the cortical neuron responds to sound in a narrow frequency range regardless of stimulus level.

nent of Doppler-shift. It is not yet known whether neurons sensitive to either the ac or dc component aggregate to form two functional subdivisions in the DSCF area. Such an arrangement would represent subdivisions for processing either fluttering or stationary targets (Fig. 6.5).

3.1.2. How Subtended Target Angle
Is Represented by the Cerebrum

The echo amplitude is the result of the target cross-sectional area, the inverse 4th power of the target distance and other factors. For simplicity, it may be stated that echo amplitude is related to the subtended target angle. Then the representation of subtended target angle by the brain is the same as the representation of echo amplitude. At the periphery, neurons show a relatively monotonic impulse-count function. That is, the number of impulses per stimulus increases monotonically with stimulus amplitude.

In the auditory cortex, however, neurons show a wide variation in impulse-count function (Fig. 6.7). They are apparently "tuned" to particular stimulus amplitudes. In a sharply tuned neuron, the peak of the impulse-count function in response to a tone burst at its best frequency uniquely indicates the best stimulus amplitude for its maximum excitation, *the best amplitude.* It might be expected that the response at best amplitude is weaker for neurons with smaller best amplitudes. This is, however, not the case. The responses at the best amplitude are usually equally strong in all neurons with different best amplitudes. It is expected, therefore, that larger numbers of excitatory inputs are integrated in neurons with smaller best amplitudes.

In the DSCF area, each orthogonal microelectrode penetration is characterized by neurons with not only nearly identical best frequencies and excitatory areas, but also nearly identical best amplitudes and impluse-count functions (31). Interestingly, the best amplitude characterizing individual cortical columns systematically varies with the location on the cortical surface, so that there is a systematic *amplitopic representation*; neurons tuned to sounds of 20–30 dB SPL are located ventrally and those tuned to strong sounds of 90–100 dB SPL are located dorsally (Fig. 6.8, parts B and D). The amplitude axis (subtended angle axis) is circular. That is, it crosses the frequency axis. *The dynamic range of the amplitopic representation is about 70 dB on the average* and *the best amplitude varies at a rate of 0.4 dB/degree* along the amplitude axis.

The *frequency vs amplitude coordinates* found in the mustached bat are the first demonstration that the amplitude spectrum of a signal is represented by a spatial pattern of neural activity. Although the coordinates have thus far been studied only for sounds

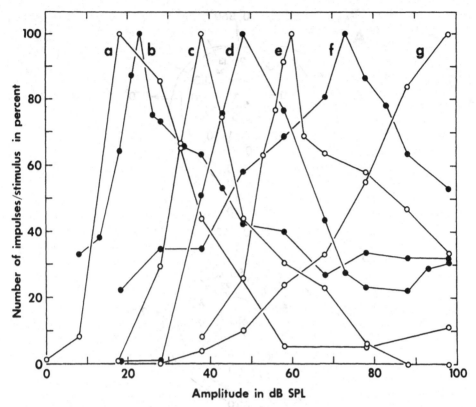

FIG. 6.7. Impulse-count functions of seven single neurons recorded in seven different penetrations within the DSCF area. Each curve represents the percentage of impulses per 50-ms-long tone burst as a function of stimulus level in dB SPL (decibels referred to 0.0002 dyne/cm^2 rms). At 100%, these neurons discharged 3–5 impulses at the onset of or during each stimulus. The frequency of the tone burst was at the best frequency of each neuron, which ranged from 61.25 to 63.97 kHz (Suga, 1977, ref. 31; by permission of the Amer. Assoc. Adv. Sci.).

in a limited frequency range, they are the first example that truly supports the spatiotemporal pattern hypothesis (see two slightly different interpretations in section 3.5)

It should be noted that the DSCF area is disproportionately large in the auditory cortex and that, furthermore, *it itself is disproportionately organized to represent acoustic signals of 61.5–62.0 kHz and 30–50 dB SPL over a much larger area.* The frequency and amplitude of echoes from prey may be predominantly in this range from the moment of target detection through the target-directed flight. Biologically important echoes are apparently over-represented by the activity of many neurons specialized for fine echo processing. As is already clear, the tonotopic and amplitopic

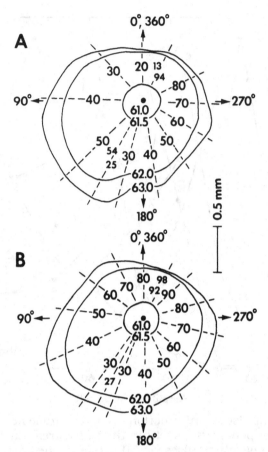

FIG. 6.8. A and B: The largest circles represent the DSCF area, in which the best frequency (BF) and best amplitude (BA) contours are shown by solid and dashed lines, respectively. The BF in kilohertz and BA in dB SPL are respectively shown by three-digit and two-digit numbers. The averaged BA contour maps were composed from the data shown in C and D. The contours for 64.0–66.0 kHz, not shown, are just outside the ventral part of the 63.0 kHz contour (see Fig. 6.5). It should be noted that the BF and BA contour maps of the individual bats are not as smooth as the average ones.

representations have certain functional significance for the mustached bat, that is, the frequency of an echo (target velocity information) is represented along the radial axis, while the amplitude, which is related to target range and size, is expressed along the circular axis (Figs. 6.5 and 6.8).

The neural mechanism for amplitopic representation is systematic variation in the extent of lateral inhibition and also in the minimum threshold for excitation. The overlap of an inhibitory area(s) with an excitatory area produces not only a sharper excita-

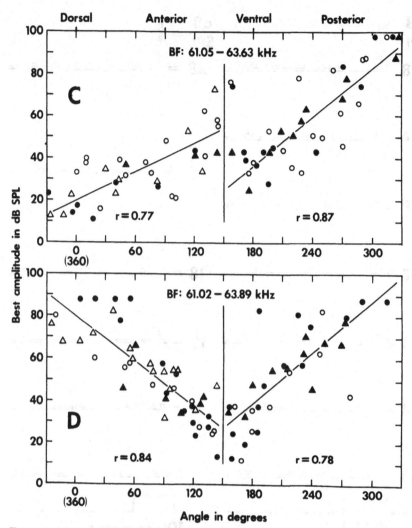

FIG. 6.8 C, D: Each of the graphs represents the data obtained from four bats, indicated by four different symbols. The ordinates and abscissas represent BAs in dB SPL and angles around the center of the area in degrees, respectively. The angle is expressed counterclockwise, starting from the dorsal part of the DSCF area. The regression coefficient (r) is shown below each regression line. The range of BFs of neurons sampled is also shown in each graph. The amplitopic representation in C and D are called N- and V-types, respectively. Thus far, six animals showed the N-type and twelve showed the V-type (Suga, 1977, ref. 31; by permission of the Amer. Assoc. Adv. Sci.).

tory area, but also a nonmonotonic impulse-count function, so that best amplitude is usually just below the threshold of the inhibitory area at the best excitatory frequency (Figs. 6.6 and 6.9; 27). The

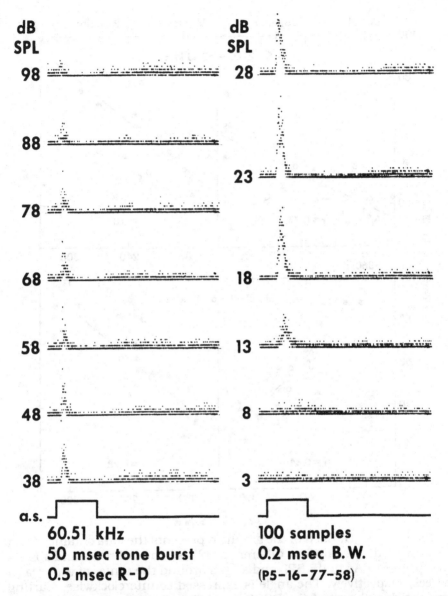

FIG. 6.9. A nonmonotonic impulse-count function of a single neuron in the DSCF area. The acoustic stimulus (a.s.) is a 50-ms-long tone burst of 60.51 kHz, which is the best frequency of this neuron. The amplitude of the tone burst is indicated by the figure on the left of each PST histogram. The minimum threshold and best amplitude of the neuron are 8 and 23 dB SPL, respectively. The histograms each represent the responses of the neuron to an identical stimulus delivered 100 times at a rate of 1.5/s. Note the clear inhibition following the on-response at stimulus levels higher than 28 dB SPL.

minimum threshold for excitation also contributes to the production of various amplitude-sensitivity curves, because it systematically changes along with best amplitude on the amplitopic axis. The dynamic range is, however, only about 40 instead of 70 dB. In other words, the slope of a nonmonotonic impulse-count function at larger stimulus amplitudes is formed by lateral inhibition and the slope at smaller stimulus amplitudes by "adjusting" a minimum threshold for excitation (31, 38).

In the auditory system, lateral inhibition plays several important roles: (a) production of *"level tolerant" bandwidths,* i.e., sharpening of tuning curves, (b) production of *amplitude selectivity,* (c) *sharpening of amplitude-spectrum representation* and (d) production of *specialized neurons* that may be considered IBP filters or feature detectors. Such important roles of lateral inhibition are facilitated by a systematic tonotopic organization of the auditory system. If the auditory system were not tonotopically organized, formation of the neural circuits necessary for these roles would be very difficult.

3.1.3. How Target Direction is Represented by the Cerebrum

The DSCF area shows an additional interesting aspect of functional organization. Almost all neurons in this area are excited by contralateral stimuli and are either excited or inhibited by ipsilateral stimuli. They are called E-E and I-E neurons, respectively. E-E neurons are poorly directional and are equally sensitive to a sound source in front of the animal between left 30° and right 30°. On the other hand, I-E neurons are directionally sensitive. Their response dramatically changes as function of azimuth angle between left 30° and right 30°. Here, "excitation" evoked by ipsilateral or contralateral stimuli means that the stimuli evoke clear excitation at least at particular stimulus amplitudes. In neurons showing nonmonotonic impulse-count function, weak excitation is usually followed by pronounced inhibition, when an excitatory stimulus is intense.

Each microelectrode penetration orthogonal to the DSCF area is characterized not only by best frequency and best amplitude, but also by either E-E neurons, I-E neurons or neurons exhibiting binaural interaction that vary with depth. Thus this area consists of at least three types of binaural columns. *E-E neurons (or columns)* are located mainly in the area corresponding to lower best amplitudes, whereas *I-E neurons (or columns)* are distributed mainly in the area corresponding to higher best amplitudes. In other words, neurons more sensitive to weaker echoes integrate (or even multiply) signals from both ears for effective target detection. Neurons more sensitive to stronger echoes, on the other hand, are suited for processing directional information about a target. Thus, the DSCF area has not only coordinates of frequency vs amplitude, but it consists of functional subdivisions, each of which is suited for either target detection or target localization (Fig. 6.5). It would be expected that, when the mustached bat finds a target, E-E neurons are first excited and that, when the bat flies toward it, the echo intensity becomes stronger, exciting neurons tuned to moderate and strong amplitudes. They are interestingly suited for processing directional information (15).

In the barn owl, *Tyto alba*, a neural map of auditory space has been found in a part of the midbrain auditory nucleus (14). It has not yet been determined whether echolocating bats also have such a neural map.

3.1.4. How Target Range is Represented by the Cerebrum

One of the most important aspects of echolocation is ranging. The primary cue for ranging is the delay of an echo from an emitted sound. Peripheral neurons respond to both the orientation sound and an echo. The interval between these grouped discharges is directly related to the echo delay. Therefore, target range is coded by the interval between the grouped discharges (Fig. 6.1, part 4b). In the *FM–FM area*, however, most neurons respond poorly or not at all to either the orientation sound alone or echo alone, but vigorously respond to orientation sound-echo pairs with particular echo delays. This area is apparently specialized for processing echo delay for target ranging.

As already described, the Panamanian mustached bat emits orientation sounds containing up to four harmonics (H_1–H_4), of which the second harmonic is always predominant and the first is usually much weaker. Each harmonic consists of CF and FM components. Therefore, there are up to eight components in each emitted signal (CF_{1-4}, FM_{1-4}). Echoes eliciting behavioral responses in the mustached bat usually overlap with the emitted signal. As a result, biosonar information must be extracted from a complex sound that potentially contains up to 16 components. The duration and repetition rate of sound emission systematically vary in a target-directed flight (Fig. 6.2A). We, therefore, built a pair of harmonic generators to synthesize such orientation sounds and echoes in the three phases of echolocation.

Neurons in the FM–FM area show maximum responses to such synthesized orientation sound and echo pairs with particular echo delays. Many signal elements can, however, be eliminated without reducing the response, because the neurons were found to be sensitive to particular combinations of two or three signal elements. Interestingly, the essential elements in such paired stimuli are found to be only the first harmonic FM component (FM_1) in the orientation sound and one or more higher harmonic FM components (FM_{2-4}) in the echo. The CF components have no significant effect on their responses. Therefore, these neurons are called *FM_1–FM_n facilitation neurons.** For the maximum excitation of some of them, the pair of FM signals must have specific relationships not only in time but also in amplitude and frequency. Thus, these neurons are tuned to a target at a particular distance, of a particular cross-sectional area and moving at a particular relative velocity.

In the FM–FM area, tonotopic representation is vague and systematic amplitopic representation has not been found, although the neurons are often tuned to particular frequency and amplitude relationships of two sounds. The FM–FM area consists of three major clusters of FM_1–FM_n facilitation neurons: FM_1–FM_3, FM_1–FM_4 and FM_1–FM_2 neurons, which are usually arranged dorsal-to-ventral in this order (Figs. 6.5 and 6.13B). The PST histograms of Fig. 6.10 illustrate the response of an FM_1–FM_2 facilitation neuron to stimulus pairs repeated at 10/s (CF, 30 ms; FM, 4 ms duration), 40/s (CF, 15 ms; FM, 3 ms) and 100/s (CF, 5 ms; FM, 2 ms). Note that the delay resulting in the best response to the stimuli (the *best delay*) at 40/s, for instance, is 6.6 ms and that there is almost no response to presentation of the orientation sound or echo components or the stimulus pair with 0.6 or 13 ms echo delay.

---

*The dash between FMs means that successive delivery of two sounds is essential for maximum excitation of combination-sensitive neurons; "n" designates the specific FM harmonics of the echo to be combined with the FM_1 of the orientation sound (n = 2,3 or 4 or their combinations). For instance, FM_1–FM_2 means that FM_2 should be delivered with a particular delay from FM_1 for best facilitation. Multiple suffixes, for instance, $FM_{2,3}$ in FM_1–$FM_{2,3}$ facilitation neurons, means that either FM_2 or FM_3 delivered after FM_1 effects the same or similar facilitation. FM_1–FM_n facilitation (or specialized) neurons are those whose response to FM_n is facilitated only by the FM_1 of H_1, but not CF_1. On the other hand, H_1–FM_n facilitation (or specialized) neurons are those whose response to FM_n is facilitated by H_1 or its components CF_1 and FM_1. The frequency of CF_1 for maximum excitation is usually slightly lower than that of orientation sounds recorded in nature and is often near the center frequency of FM_1 (e.g., Fig. 6.16). In the natural situation, therefore, H_1–FM_n-facilitation neurons would not be facilitated by the combination of CF_1 and FM_n, but the combination of FM_1 and FM_n as FM_1–FM_n facilitation neurons would. Therefore, H_1–FM_n facilitation neurons are called FM_1–FM_n facilitation neurons, when their function is only described as in this article. This simplification would help understanding of the story. When the description about these neurons is concerned with the neural mechanisms for production of their response properties, however, the term H_1–FM_n facilitation neurons, would be used (e.g., Fig. 6.16).

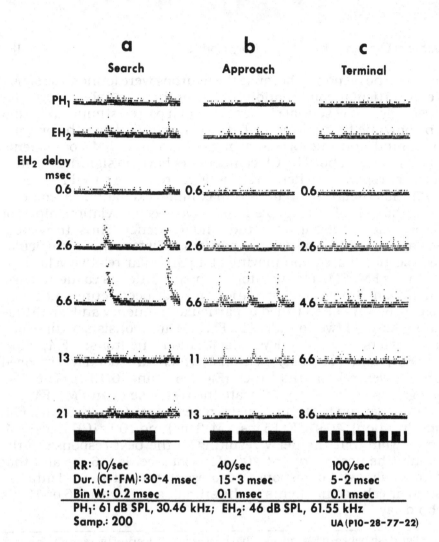

FIG. 6.10. PST histograms of the responses of a tracking neuron (FM$_1$–FM$_2$ facilitation) to different echo delays during simulated search, approach and terminal phases of target-directed flight. The essential harmonics for facilitation of this neuron were H$_1$ of the orientation sound (CF$_1$, 30.46 kHz) and H$_2$ of the echo (CF$_2$, 61.55 kHz). The amplitudes of these harmonics (PH$_1$ and EH$_2$) were fixed at 61 and 46 dB SPL, respectively. The upper two histograms in each column show the poor response or lack of response of the neuron to PH$_1$ or EH$_2$ delivered alone at each repetition rate. The five histograms below them show the facilitation at various echo delays, when PH$_1$ and EH$_2$ were delivered as a pair. Note the decrease in the range of delays, over which the neuron followed the stimulus pair, and also the shortening of the best delay as the repetition rate increased from 10 to 100 pairs/s (corresponding to a change in distance from 114 to 55 cm). Stimulus markers beneath the histograms indicate the orientation sound of the stimulus pair only. The data were obtained from an unanesthetized bat (O'Neill and Suga, 1979, ref. 19; by permission of the Amer. Assoc. Adv. Sci.).

In some neurons the best delay changed as a function of repetition rate and duration of stimuli, while it remained relatively constant in others. We call these two types *tracking* and *range-tuned neurons*, respectively. Their response properties are better illustrated by measuring their *delay-tuning curves* (Fig. 6.11). These curves are obtained by holding the amplitude of the orientation sound or its first harmonic (PH_1) constant at the best facilitation amplitude of the individual neuron (71 ± 7 dB,$n = 113$) and measuring the threshold of response to the echo or the certain echo harmonic (EH_{2-4}) as a function of echo delay.

Range-tuned neurons have nearly the same delay-tuning curve regardless of wide variations in stimulus repetition rate (10–100/s) and duration (34–7 ms). In the neuron shown in Fig. 6.11A, the best delay (at the lowest threshold, 47 dB SPL) is 2.7 ms, corresponding to a 46 cm target range. Since a target would be at this range only during the late approach phase, this neuron is apparently specialized to respond to the echo only around that time. In addition, it is tuned to an echo amplitude of about 60 dB SPL (shaded area). Some range-tuned neurons are sharply tuned to a particular weak echo, so that their delay-tuning curves show upper thresholds (Fig. 6.11B). These neurons are thus selective for both target range and size. The slope of the leading edge of the delay-tuning curve is steep, 24 dB/ms in Fig. 6.11A and infinite in Fig. 6.11B. The width of the delay tuning curve at 10 dB above the minimum threshold is 2.8 ms for Fig. 6.11A. The tuning curve does not appear to be terribly sharp, because our criterion for threshold was very low, i.e., about 0.1 response per stimulus or 10–20% increase of background discharges (just-noticeable response). If the criterion for threshold is raised to 0.5 response per stimulus, or 50% increase from background discharge rate, the delay-tuning curve becomes much narrower and its minimum threshold slightly higher. For the neurons shown in Fig. 6.11A and B, for instance, the delay-tuning curve would become a small circle surrounding the shaded area.

The most appropriate way to express the response properties of range-tuned neurons is the isoimpulse-count contour map plotted on the coordinates of echo amplitude vs echo delay. The contours usually appear as ellipsoids around the intersection of particular best delay and best facilitation amplitude, indicating that the neurons are tuned in both echo delay and amplitude (e.g., Fig. 6.12). That is, *they are tuned to targets with particular cross-sectional areas at specific ranges. For their excitation, a Doppler-shift is also an influential parameter.*

For maximum facilitation of range-tuned neurons, the amplitude of the FM_1 of the orientation sound usually ranges from 60 to 80 dB SPL, while that of the FM_n of the echo is more variable, a

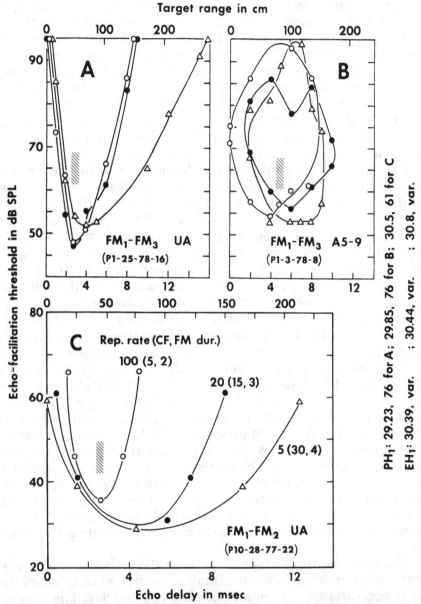

FIG. 6.11. Delay-tuning curves for three neurons in the FM–FM area. A, Range-tuned neuron (FM₁-FM₃ facilitation), recorded from an unanesthetized bat. The delay-tuning curves for the search (△, 10 pairs/s) mid-approach (●, 40/s) and terminal (○, 100/s) phases are very sharp and nearly identical for the approach and terminal phases. The best delay is also the same, 2.7 ms. This neuron is tuned to a target 46 cm away during the late approach to terminal phase. The best echo amplitude for facilitation is about 65 dB SPL. B, Another range-tuned neuron (FM₁–FM₃ facilitation) demonstrating upper thresholds for facilitation. The three curves

single value (best facilitation amplitude) between 20 and 80 SPL or any value over a wide range. Facilitation usually does not occur or is very poor, when echo FM_n is stronger than 80 dB SPL. These neurons fulfill the requirements of the biosonar system, since the amplitude of vocal self-stimulation by emitted orientation sounds is relatively constant, while echo amplitude is quite variable. The *best delay* (i.e., *best range*) and best facilitation amplitude are different from neuron to neuron. A series of such neural filters could process range information and also probably could characterize targets (Fig. 6.5; see section 3.5).

The obvious question is whether these range-tuned neurons with different best delays (best ranges) are systematically arranged along an axis representing target range information. The FM–FM area shows columnar organization; i.e., neurons arranged orthogonally to the surface are characterized by nearly identical response properties, including best delays. For instance, when a neuron recorded near the surface showed a narrow delay-tuning curve with a best delay of 4.5 ms all other neurons recorded in the same penetration also showed similar narrow delay-tuning curves with a 4.5-ms best delay. Confirmation of the columnar organization of best delays simplified our study of cortical representation of target range, because we could rely on the uniformity of activity at different depths in the cortex.

FIG. 6.11. (continued)
are similar and the neuron is tuned to a target that returns an echo of about 65 dB SPL from a distance of about 85 cm. The data were obtained from a lightly anesthetized bat between 5 and 9 h after Nembutal administration (O'Neill and Suga, 1979, ref. 19; by permission of the Amer. Assoc. Adv. Sci.). C, A tracking neuron (FM_1–FM_2 facilitation) shows broad delay-tuning with echo facilitation thresholds of 29–30 dB SPL and best delays of 3.8–5.0 ms (target range, 65–85 cm) for search and early approach phases ($\triangle$, 5/s; ●, 20/s: these data were taken at an early stage of the research before repetition rates were standardized). During the terminal phase (○, 100/s, the best delay shortens dramatically to about 2.5 ms (43 cm distance), the threshold increases to about 36 dB SPL and the delay-tuning curve becomes much narrower. The data were obtained from an unanesthetized (UA) bat. In A and B, each stimulus in a pair had three harmonics, as shown in Fig. 6.2A, while in C, only the H_1 of the orientation sound and the H_2 of the echo were paired, because other harmonics had no influence on the facilitation response of this neuron. The frequencies and amplitudes of the CF_1 components of the orientation sounds (PH_1) and the frequencies of the CF_1 components of the echoes (EH_1) used for the measurements are shown at the right. The shaded area in each graph indicates the best echo amplitude for facilitation in the terminal phase. (Suga, O'Neill and Manabe, 1978, ref. 41; by permission of the Amer. Assoc. Adv. Sci.).

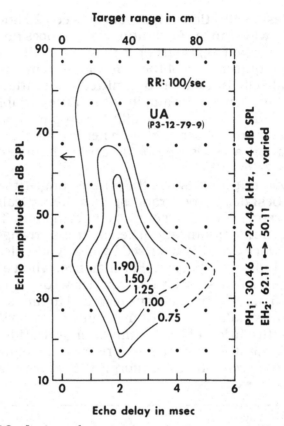

FIG. 6.12. Isoimpulse-count contours representing the response magnitude of a range-tuned neuron (FM_1–FM_2 facilitation) plotted on the coordinates of echo amplitude vs delay (or target range). The outermost contour line (0.75 impulse per paired stimulus) represents the delay-tuning curve, when the definition of threshold is an echo evoking just-noticeable response (about 20% increase from background discharge rate). When the definition is changed to, for instance, 1.0 impulse per paired stimulus, the delay-tuning curve shrinks as indicated by the 1.0 line. This neuron is tuned to a target that is located 34 cm in front of the bat and that returns an echo of 37 dB SPL to the bat's ears. Since the neuron was tuned to targets at a short distance, the orientation sound–echo pair for this plot was delivered at a rate of 100/s mimicking the terminal phase of echolocation (see Fig. 6.2A). The essential components for excitation were the H_1 of the orientation sound (PH_1) and the H_2 of the echo (EH_2). The former consisted of a 5-ms CF of 30.46 kHz followed by a 2-ms FM sweeping from 30.46 to 24.46 kHz, at an amplitude of 64 dB SPL (arrow). The latter consisted of a 5-ms CF of 62.11 kHz followed by a 2-ms FM sweeping from 62.11 to 50.11 kHz, at an amplitude that was varied in increments of 10 dB. Note that, for the maximum response, the echo H_2 should be Doppler-shifted by 1.19 kHz. This Doppler-shift could be generated by the relative motion of a target. Individual dots indicate where an average number of impulses per paired stimulus was obtained by pres-

During an oblique microelectrode penetration through the FM–FM area in the rostrocaudal direction, best delay becomes systematically longer. Figure 6.13B summarizes the results and represents isobest-delay contour lines comprising a target range axis. Neurons with extremely short best delays are always located only in the rostral part of the FM–FM area. The shortest best delay of a range-tuned neuron obtained thus far is 0.4 ms corresponding to a target range of 6.9 cm. Neurons with a very short best delay responded strongly by discharging action potentials in a tight cluster to each paired stimulus delivered at a rate of 100/s, mimicking the terminal phase of echolocation. When best delay is shorter than 2.0 ms, the FM component (2.0 ms long) of the echo in the terminal phase overlaps that of the orientation sound in our experiments. The delay tuning curve for such an extremely short best delay is very sharp, but nevertheless often crosses the 0 ms delay line at 60–80 dB SPL. Therefore, when the orientation sound is delivered at greater than 60 dB SPL, facilitation is evoked by the combination of different harmonics in the orientation sound *per se* and is further augmented by an echo with a very short delay. Response latencies of range-tuned neurons to the echo FM are short, 7–10 ms. These findings clearly indicate that *the auditory cortex is involved in information processing even in the terminal phase*. Data comparable to the above have also been obtained from the little brown bat, *Myotis lucifugus*, which uses FM orientation sounds (W.E. Sullivan, unpublished). Therefore, the old idea that information processing in the terminal phase is performed only in the subcollicular auditory nuclei should be revised.

Neurons with long best delays are located in the caudal part of the FM–FM area (Fig.6.13B). The longest best delay obtained thus far is 18 ms, corresponding to a target range of 310 cm. The delay-tuning curves of neurons with best delays longer than 12 ms are broad and they respond strongly to each paired stimulus only when delivered at lower repetition rates characteristic of the search phase. The response is very poor at a rate of 40/s and completely disappears at 100/s. Their role in range discrimination at distances longer than 200 cm may be limited, because of their broad delay-

Fig. 6.12. (continued)
enting the identical paired stimulus 200 times. The contour lines are drawn on the basis of these data points. The dashed parts of the contour lines indicate where the following of the response to each paired stimulus was not reliably observed owing to background noise associated with animal movement. The bat was not anesthetized (Suga and O'Neill, 1979, ref. 39; by permission of the Amer. Assoc. Adv. Sci.).

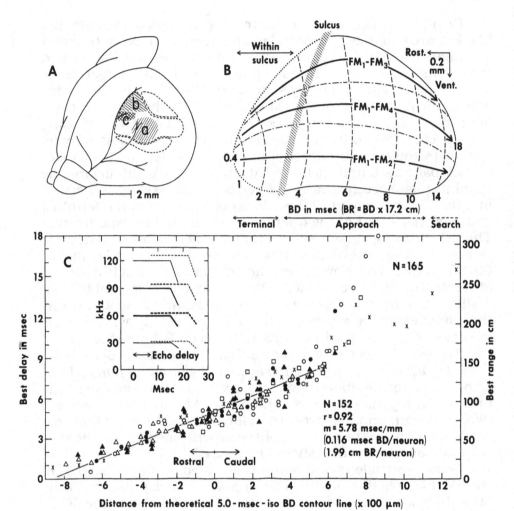

FIG. 6.13. A, The left cerebral hemisphere of the mustached bat. The large auditory cortex contains three areas specialized for processing biosonar information: a, DSCF, b, FM-FM and c, CF/CF areas. The branched lines are arteries. There is a sulcus below the largest branch. B, The FM-FM area consists of three major clusters of delay-sensitive neurons: FM_1–FM_2, FM_1–FM_3 and FM_1–FM_4 facilitation neurons. Each cluster shows odotopic representation. Iso-best-delay contours and range axes are schematically shown by dashed lines and solid arrows, respectively. Best delays (BDs) of 0.4 and 18 ms correspond to best ranges (BRs) of 7 and 310 cm, respectively. Range information in the search, approach and terminal phases of echolocation is represented by activity of different locations on the cerebral hemisphere. C, The relationship between best delay (best range) and distance along the cortical surface. The data were obtained from six cerebral hemispheres, indicated by six different sym-

tuning curves. The population of neurons with such long best delays is small.

The central part of the FM–FM area is occupied by neurons with best delays between 4 and 7 ms. Their delay-tuning curves are sharp and their responses are strong and clearly locked to each paired stimulus even at a rate of 100/s. Neurons with best delays from 3 to 8 ms are distributed over a disproportionately large area. This suggests that processing of echoes from targets from 50 to 140 cm away, i.e., information received during the approach phase, is particularly important for the mustached bat.

One quantitative expression of the neural range axis is the regression of best delays on distance along the cortical surface. Therefore, best delays are plotted as a function of distance from the 5.0-ms iso-best-delay contour line (Fig.6.13C). The correlation coefficient for the best delays between 0 and 10 ms is 0.92 (N = 152). The slope of the regression line is 5.78-ms best delay/mm of cortical surface. Since the average interneuronal distance in the cortical plane measured with the frozen sections of the brain is about 20 μm, *successive neurons can express target range in 1.99 cm increments.*

These neurophysiological data show an interesting correspondence with some available behavioral data. The little brown bat, *Myotis lucifugus*, begins the approach phase to 0.3-cm diameter wires at an average distance of 225 cm (11) and the horseshoe bat, *Rhinolophus ferrumequinum*, compensates for Doppler-shifted echoes only when delayed less than an average of 17.5 ms (301 cm) (21). That bats react to targets at distances shorter than 301 cm by increasing the rate of sound emission corresponds to our finding that the range axis ends at about 310 cm. [Many species of bats probably detect large targets at distances much farther than 300 cm (4)]. The bats, *Eptesicus fuscus*, *Phyllostomus hastatus*, *Pteronotus suapurensis* and *R. ferrumequinum*, are able to discriminate 1.2–2.5 cm range differences at an absolute distance of

---

FIG. 6.13. (continued)
bols. The regression line represents an average change in best delay with distance along the cortical surface. Since the 5-ms iso-BD contour line always crossed the central part of the FM–FM area along the exposed surface of the cortex, the 5-ms BD on the regression line is used as a reference point to express distance. The inset in C is a schematized sonagram of an orientation sound and a Doppler-shifted echo in the approach phase of echolocation. All the data were obtained from unanesthetized bats (Suga and O'Neill, 1979, ref. 39; by permission of the Amer. Assoc. Adv. Sci.).

30 and/or 60 cm (22).* This would also correspond to our finding, if we assume that the 1.99 cm increment per neuron is the theoretical limit of just-noticeable difference in distance.

Our data are the first demonstration that echo delay, i.e., target range, is represented by excitation of range-tuned neurons systematically arranged according to their best delays along an axis in the central auditory system. We thereby name this organization *odotopic representation* (39). It should be noted that this odotopic representation is based upon the best delays of range-tuned neurons, so that the representation is the same regardless of large variations in repetition rate (10–100/s) and duration of signals (7–34 ms). The synthesis of a range axis, which has no anatomical precursor at the periphery, is suggestive of the methods by which sensory information may be extracted and represented in the brain.

In contrast to range-tuned neurons, the other type of delay-sensitive neuron, a tracking neuron, changes both its best delay and the shape of the delay-tuning curve as the repetition rate and duration of signals change (Fig.6.11C). These neurons respond to targets over a broad range during the search and early approach phases. But during the late approach and terminal phases, the best delay becomes shorter and the delay-tuning curve narrower. Thus, these neurons can respond to the echo during all phases of target pursuit. The most interesting feature of these neurons is the narrowing of the delay tuning curve as the bat approaches the target, so that the neuron effectively "ignores" possible secondary echo sources from objects farther away.

It is puzzling that the first harmonic (H_1, in particular FM_1) of the orientation sound is so critical to the response of range-tuned and tracking neurons in spite of the fact that H_1 is much weaker than the second and third harmonics and sometimes is hardly detectable in recordings in confined spaces (our laboratory). This , however, may be a fascinating adaptation of the bat's auditory system to reduce the probability of jamming by the sounds of echolocating conspecifics. When we first delivered pairs of orientation sounds and echoes without H_1, we found no range-sensitive neurons in the FM–FM area. This means that range-sensitive neurons are protected to some extent from a jamming effect of orientation sounds and/or echoes produced by conspecifics flying nearby in a small space. To excite range-sensitive neurons, H_1 must stimulate the ears prior to an echo, in spite of its weakness in the

---

*Simmons (23) found that the big brown bat, *Eptesicus fuscus*, can detect jitter in echo delay as small as 0.5 μs and concluded that peripheral auditory neurons code phase information in the form of phase-locked discharges that would occur even for a 100 kHz sound. The neurophysiological findings thus far available are not appropriate for comparison with his recent data.

emitted sound. This suggests that H_1 produced by the vocal cords stimulate the animal's own ears by bone conduction, while it is not emitted at a significant amplitude possibly because of suppression by vocal-tract antiresonance. *In nature, range-sensitive neurons would be selectively excited, when the animal itself emits orientation sounds and echoes return with particular delays.* Without this property, target ranging in a small space would be frequently impaired by conspecifics. Thus, these neurons are jamming-tolerant in most situations.

Behavioral experiments on range discrimination clearly indicate that "CF–FM" bats use the FM component for ranging just as "FM" bats do (22, 24). The essential components in the orientation sound and echo for the excitation of delay-sensitive neurons, which we have found, are certain combinations of FM components, so that our neurophysiological data parallel the behavioral results. Further, delay-sensitive neurons, apparently of this type, have been found in the intercollicular nucleus of an FM bat, *Eptesicus fuscus* (2), and in the auditory cortex of another FM bat, *Myotis lucifugus* (W.E. Sullivan, unpublished). The importance of FM signals in ranging in both CF–FM and FM bats is thus demonstrated neurophysiologically.

In order to produce the response properties of range-tuned or tracking neurons, lower-order neurons with a broad spectrum of recovery cycles (5, 44) and facilitation to echoes (9, 10) are necessary. The neural network model for ranging, proposed by Suga and Schlegel (44), consists of three neural levels, (i) tonic neurons with short recovery cycles, (ii) phasic latency-constant neurons with a broad spectrum of recovery cycles and (iii) neurons with different time gates (slots) for echo detection and ranging. Modifying this model in the light of our cortical data, the second level should contain neurons with a recovery cycle, which includes a facilitation period preceded and/or followed by an inhibitory period, and the third level should consist of neurons showing facilitation to the orientation sound-echo pair without a response to the orientation sound or echo alone. At the third level, range-tuned neurons express the extent to which echo delays match their best delays by the magnitude of the excitation. It is not necessary for this type of neuron to be very phasic and latency-constant, like those at the second level. The model contains the minimum of neural elements necessary for ranging, which are supported by neurophysiological data. It is, of course, subject to further elaboration with the light of future results. Neurons, which respond to both the orientation sound and echo even at short echo delays, are important neuronal elements for ranging, but represent the lowest level of neurophysiological activity in that they simply reflect the tonic responses of peripheral neu-

rons. The auditory system undoubtedly processes range information by more elegant means than this. This does not mean that
discharges synchronized with an emitted sound and an echo are
not concerned with range perception. Synchronous discharges
may take a complementary role in range perception, in particular,
when a target range is longer than 2 m.

3.1.5. Complex-Sound Processing Area
in the Cerebrum

As described above, neurons in the FM–FM area are sensitive to
particular combinations of FM components in complex sounds.
Ventral to this is the *CF/CF area*, in which neurons are, on the contrary, sensitive to particular combinations of CF components in
complex sounds. In other words, *neurons in these two areas "assemble" particular signal elements* or *"examine" combinations of
particular signal elements for processing complex sounds*. Accordingly, these two areas may be called the *complex-sound processing area. The FM–FM area demonstrates how a sequence of
sounds is processed in the central auditory system*, while *the
CF/CF area*, as described below, *demonstrates how overtone structure is processed.*

In the CF/CF area, neurons show remarkable facilitation, when
the CF component of the first harmonic (CF_1) is simultaneously delivered with one or more CF components of higher harmonics
(CF_{1-4}). There are two types of CF_1/CF_n *facilitation neurons:*
CF_1/CF_2 and CF_1/CF_3. The slash means that the harmonic components must be delivered *simultaneously*, or with overlap to elicit
the best facilitation. CF_1/CF_2 facilitation neurons form one cluster,
which is dorsal to the cluster of CF_1/CF_3 in most of the brains studied (Fig.6.5; 42). Almost all of them respond to one or both individual components delivered alone at amplitudes greater than 80 dB
SPL. These responses are, however, often not prominent or consistent and the latency is usually long and fluctuating. When two or
more components are combined in an appropriate frequency, amplitude and temporal relationship, however, responses become
prominent and consistent and the response latency is shorter and
more constant. Figure 6.14A demonstrates facilitation of responses of a CF_1/CF_2 neuron by PST histograms. This neuron
showed some response to CF_1 alone, but no response to CF_2 alone.
When CF_1 and CF_2 were delivered simultaneously, however, the
neuron responded strongly. The response latency was short,
6–8ms.

The principle, that the larger the amplitudes of combined
components, the larger the facilitation, does not hold. Instead,
maximum facilitation is usually evoked only when two or three
components are delivered in a particular amplitude relationship

unique to the individual neurons (e.g., Fig.6.14). An impulse-count function, which relates stimulus amplitude to the number of impulses per stimulus, clearly demonstrates this point. In Fig.6.14B, for instance, CF_2 should be 56 dB SPL and CF_1 should be between 50 and 70 dB SPL for maximum facilitation. Compared with CF_1 alone, the threshold for CF_1 with CF_2 is only 8 dB lower, but the magnitude of response (number of impulses per stimulus) is nearly two times larger. Compared with CF_2 alone, the threshold for CF_2 with CF_1 is 53 dB lower and the response magnitude is four times larger. According to our definition, facilitation always involves an increase in number of impulses, while it is not necessarily associated with a decrease in threshold.

Facilitation properties vary among neurons: facilitation with a dramatic decrease in threshold, facilitation without a decrease in threshold, etc. Examples of excitatory-frequency-tuning curves and *facilitation-frequency-tuning curves* of four single neurons are presented in Fig.6.15. A CF_1/CF_2 facilitation neuron in Fig.6.15A responds poorly to single tones, with minimum thresholds of 58 dB SPL at 28.91 kHz and 74 dB SPL at 60.55 kHz. However, it responds vigorously to CF_1/CF_2, with minmum thresholds of 31 dB SPL at 29.75 kHz and 37 dB SPL at 59.39 kHz. It should be noted that the thresholds for both CF_1 and CF_2 are lowered by facilitation and that the best frequencies of CF_1 and CF_2 are harmonically related. A CF_1/CF_3 facilitation neuron in Fig.6.15B responds poorly to single tones, with minimum thresholds of 76 dB SPL at 29.10 kHz and 43 dB SPL at 91.53 kHz. (In spite of the low threshold for a 91.53 kHz tone, the response to it is poor and fluctuates widely.) The neuron responds strongly to CF_1/CF_3, with minimum thresholds of 38 dB SPL at 28.91 kHz and 34 dB SPL at 91.53 kHz. The decrease in threshold by facilitation is prominent for CF_1 but small for CF_3. In Fig. 6.15C, a CF_1/CF_3 facilitation neuron is shown that exhibits different properties from those in Fig.6.15B. Its single-tone threshold and facilitation threshold are the same for CF_1, but different by 65 dB for CF_3. In both neurons shown in 15B and C, the best frequencies for facilitation are quasiharmonically related.

CF_1/CF_2 facilitation neurons are found in a large cluster usually dorsal to a cluster of CF_1/CF_3 facilitation neurons. Between these two clusters there are $CF_1/CF_{2,3}$ facilitation neurons forming either an independent cluster or a transitional area between clusters. Figure 6.15D represents the excitatory frequency-tuning and facilitation-frequency-tuning curves of a $CF_1/CF_{2,3}$ facilitation neuron. The response of this neuron is facilitated, when CF_1 is delivered together with CF_2 or CF_3. The thresholds for CF_1, CF_2 and CF_3, respectively, decrease by 14, 27 and 73 dB with the facilitation. CF_2/CF_3 evokes no facilitation. The response is largest

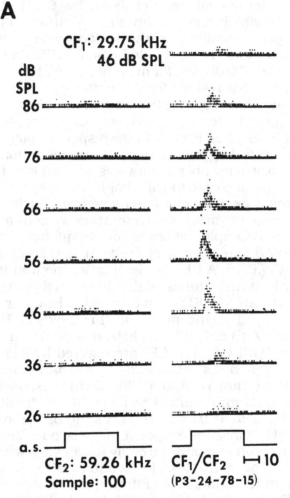

FIG. 6.14A. PST histograms of responses of a CF_1/CF_2 facilitation neuron. The response to the CF_1 alone is shown at the top of the right column and the responses to the CF_2 alone and CF_2 with CF_1 are shown in the left and right columns, respectively. The acoustic simuli (a.s.) are 34 ms in duration and 0.5 ms in rise-decay time. The CF_1 is 29.75 kHz and 46 dB SPL, while the CF_2 is 59.26 kHz and is either 26, 36, 46, 56, 66, 76 or 86 dB SPL. Each PST histogram consists of neural activity for 100 presentations of the same sound(s).

when these three tone bursts are delivered simultaneously at amplitudes indicated by the three points marked × in Fig. 6.15D. The best frequencies for facilitation, 30.13, 60.63 and 90.26 kHz, are harmonically related.

The harmonic relationship among CF best frequencies for facilitation was statistically examined after normalizing to the mean

B

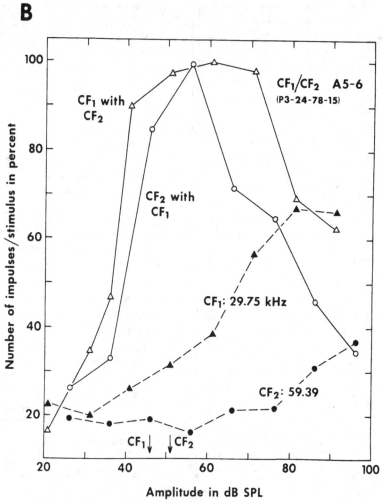

FIG. 6.14B. Impulse-count functions of the CF_1 alone, CF_2 alone, CF_1 with CF_2 of 51 dB SPL and CF_2 with CF_1 of 46 dB SPL. The number of impulses/stimulus was counted for 200 ms after the onset of the 34 ms-long tone burst delivered 100 times. The 200-ms period was long enough to include any possible change in discharge rate evoked by the stimulus. The data were obtained from a lightly anesthetized bat between 5 and 6 h after Nembutal administration (Suga, O'Neill and Manabe, 1979, ref. 42; by permission of the Amer. Assoc. Adv. Sci.).

best frequency of CF_1 29.80 ± 0.80 kHz (N = 372). The mean and standard deviations are 60.55 ± 0.98 kHz for CF_2 (N = 204) and 91.51 ± 1.04 kHz for CF_3 (N = 205). For maximum facilitation, a precise harmonic relationship between two or three CF signals is required for some neurons, but instead, particular amounts of Doppler shift are required for others. The best facilitation fre-

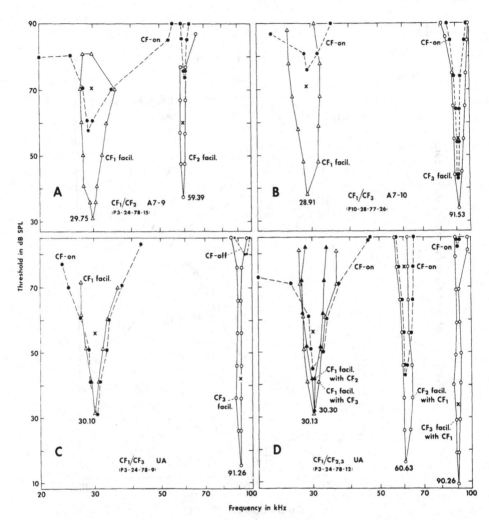

Fig. 6.15. Excitatory (dashed) and facilitation (solid) areas, i.e., the areas above or surrounded by excitatory-frequency-tuning or facilitation-frequency-tuning curves of four neurons. An excitatory area was measured by delivering a 34 ms-long CF tone, while a facilitation area was measured by delivering a 34 ms-long CF tone (conditioning tone) simultaneously with another 34 ms-long CF tone (test tone) at a fixed frequency and amplitude. A, A CF_1/CF_2 facilitation neuron recorded from a lightly anesthetized bat between 7 and 9 h after Nembutal administration. The test sound used to measure the CF_1 facilitation area was 59.39 kHz and 60 dB SPL, while for the CF_2 facilitation area, it was 29.75 kHz and 70 dB SPL. These are indicated by the crosses in the figure. B, A CF_1/CF_3 facilitation neuron recorded between 7 and 10 h after Nembutal administration. C, A CF_1/CF_3 facilitation neuron obtained from an unanesthetized animal. D, A $CF_1/CF_{2,3}$ facilitation neuron obtained from an unanesthetized animal. Test sounds used for the data in B, C and D are indicated by the crosses like those in A (Suga, O'Neill and Manabe, 1979, ref. 42; by permission of the Amer. Assoc. Adv. Sci.).

quency of CF_1 within one standard deviation, is 29.00—30.60 kHz. That is, most CF_1/CF_n facilitation neurons are tuned to CF_1 components of orientation sounds slightly lower than the resting frequencies. In contrast to FM_1–FM_n facilitation neurons, CF_1/CF_n facilitation neurons are best facilitated, when two or three CF signals are delivered simultaneously. Their delay-tuning curves measured with essential CF signals are broadly tuned at 0 ms. Therefore, these neurons are not sensitive to target range in the sense that FM_1–FM_n facilitation neurons are. Most CF_1/CF_n neurons do not follow well to paired components over about 40/s, although some do so up to 100 pairs/s.

The importance of CF_1/CF_n facilitation neurons to the bat is not immediately clear. One of their functions may be to process velocity information during Doppler-shift compensation (53, 54). That is, the neurons stimulated by the CF_1 of self-vocalized sounds show facilitation responses only to echoes with particular Doppler-shifted CF components. If this interpretation is correct, they would be jamming tolerant, as FM_1–FM_n facilitation neurons. The other function may be to detect the presence of nearby echolocating conspecifics that may be competing for insect prey in the same area. This would be only possible if the mustached bat enhances the first harmonic of the orientation sound, while flying in the open. On the contrary, it emphasizes the higher harmonics in a confined environment, as found in the long-eared bat, *Plecotus phyllotis* (25). The first harmonic produced by conspecifics in the open may be detected at amplitudes appropriate for facilitation, because atmospheric attenuation is less for lower frequencies. CF signals produced by conspecifics would be perceived at longer distances than FM signals, because of the high concentration of sound energy at discrete frequencies. CF_1/CF_n facilitation neurons may, thus, function in detection of nearby echolocating conspecifics.

As demonstrated in Figs. 6.14 and 6.15, CF_1/CF_n facilitation neurons are tuned to particular combinations of CF sounds. The best frequencies and best amplitudes of sounds for facilitation differ from neuron to neuron. Such variation is probably important for the detection of targets at different relative velocities and subtended angles and/or the detection of conspecific bats that emit orientation sounds differing in amplitude spectrum.

During echolocation, the detection of a bat's own Doppler-shifted echoes and sounds of other bats may be masked by self-vocalized sounds. Neurons with best frequencies near the CF components (in particular CF_2) of the orientation sound are very sharply tuned (Fig. 6.15), so that the masking effect of vocal self-stimulation would be greatly reduced by any difference in amplitude spectrum at the two ears between self-vocalized orientation sounds and those externally originating sounds. Muscular (12, 35) and/or neural (43, 46) mechanisms for attenuation of vocal self-stimulation would also play a role in reducing the masking effect.

Many species of microchiropterans are colonial. Hundreds, or even thousands of bats roost in a single cave. They are frequently found in narrow elongated caves and culverts, where they fly in opposite directions without colliding. One of the important problems in echolocation is how the bats reduce the jamming effect of biosonar sounds produced by conspecifics or how the neural representation of biosonar information is protected from jamming. We can enumerate the following seven mechanisms responsible for the reduction of jamming: (i) the sharp directionality of the orientation sound, (ii) the sharp directional sensitivity of the ear and binaural hearing, (iii) the sequential processing of echoes, (iv) the signature (subtle difference) of orientation sounds used by individual bats, (v) the "auditory" time gate, (vi) the "heteroharmonic combination", and (vii) the efferent copy. All these mechanisms would work together for successful echolocation. Some of these mechanisms have been physiologically examined, but others have not yet been studied. I would like to point out here that, different from neurons in the DSCF area, those in the FM–FM and CF/CF areas have the auditory time gate and heteroharmonic combination. These two mechanisms work for sampling the bat's own echoes and for excluding sounds produced by conspecifics (39, 54).

3.2. Extraction of Information-Bearing Elements and Examination of Their Combinations

In English speech sounds, there are several types of acoustic cues or information-bearing elements important for speech recognition. Formants (CF components) are essential for the recognition of vowels. Fills (noise bursts: NBs) are important for recognition of some fricative consonants. Transitions (FM components) and voice-onset-time (VOT) are important for the recognition of plosive consonants and some fricative consonants combined with vowels. There are also FM components in other phonemes (glides) and dipthongs. CF, FM and NB elements are found in sounds produced by many different species of animals as well. In bats, these signals are also used for communication and/or echolocation. Then, the question can be raised, whether complex sounds are processed by the auditory system employing neural circuits to extract individual types of information-bearing elements and also circuits to examine their combinations.

In the central auditory system of the bats, *Myotis lucifugus* (30) and *Pteronotus parnellii rubiginosus* (31, 32) neurons have been found that selectively respond to only one of the three types of information-bearing elements. *CF-specialized neurons* have a very narrow excitatory area sandwiched between inhibitory areas and respond only to CF tones that stimulate this narrow excitatory area. The neurons do not respond to FM sounds sweeping across the inhibitory and excitatory areas nor to noise bursts that stimulate simultaneously both the areas. Such response properties of CF-specialized neurons are explained by a neural network model producing *forward lateral inhibition*. *FM-specialized neurons* have a large inhibitory area and selectively respond to FM sounds sweeping across this area. Direction, range, speed and/or form of the frequency sweep are important for their excitation (28). They fail to show excitatory responses to CF tones and noise bursts. Such response properties of FM-specialized neurons are explained by neural network models producing *disinhibition*. *NB-specialized neurons* have a large inhibitory area and respond to noise bursts, but to neither CF tones nor FM sounds. Response properties of NB-specialized neurons must be further studied (29, 30).*

The auditory system contains neural circuits for extraction of individual information-bearing elements from complex sounds. Then, the next question is whether the auditory system also contains neural circuits that "assemble" information-bearing elements or "examine" their combinations for further processing complex sounds. The data obtained from the FM–FM and CF/CF areas demonstrate unequivocally that combinations of signal elements are examined by neurons located in separate parts of the brain (19, 39–42) and that response properties of many neurons examining combinations of signal elements result from interaction among neurons specialized for extraction of information-bearing elements (41; Suga et al., in preparation).

The histograms of Fig. 6.16, A and B represent responses of a H_1–FM_2 facilitation neuron, which shows facilitation not only for the combination of FM_1 and FM_2 but also for the combination of

---

*It is still often suggested that NB-specialized neurons and neurons selectively sensitive to *upward-sweeping* FM sounds have no biological significance, because they have been found in the central auditory system of bats, which emit only *downward-sweeping* FM sounds for echolocation. However, we know that such bats also produce noise bursts and upward-sweeping FM sounds for communication (3). Electrical stimulation of different parts of the brain elicits all of these types of sounds, including CF tones with several harmonics (45). Futhermore, it has been proposed that neurons selectively sensitive to upward-sweeping FM sounds are necessary to produce FM-specialized neurons, which selectively respond to downward-sweeping FM sounds (30,32). The significance in signal processing of some neurons found in the central auditory system may not be immediately clear. Their significance, however, may become clear, when our understanding of the communication and/or echolocation systems is deepened.

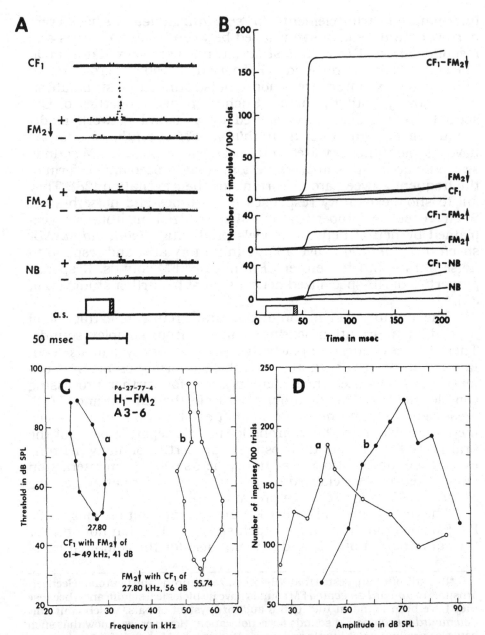

FIG. 6.16. Response properties of a single H_1–FM_2 facilitation neuron in the FM–FM area. A, PST histograms of responses. The acoustic
stimulus (a.s.) is either a 30-ms-long CF_1 alone (open rectangle) or a 4-
ms-long sound alone (shaded rectangle), which is either a downward-
sweeping FM_2 ($FM_2 \downarrow$, −) or an upward sweeping FM_2 ($FM_2 \uparrow$, −) or noise
burst (NB, −) or one of the 4-ms sounds preceded by the CF_1 (+). The CF_1
is 56 dB SPL and 27.80 kHz, which is the center frequency of the typical

CF_1 and FM_2 or H_1 and FM_2 (41; see the footnote in section 3.1.4). This neuron shows remarkable facilitation for the combination of CF_1–$FM_2 \downarrow$ (downward-sweeping FM_2), but almost no facilitation for CF_1–$FM_2 \uparrow$ (upward-sweeping FM_2) and CF_1–NB (noise burst). $FM_2 \uparrow$ and $FM_2 \downarrow$ are the same in amplitude spectrum. The NB is produced by mixing frequencies in the $FM_2 \downarrow$, so that it is comparable to $FM_2 \downarrow$ in amplitude spectrum. The response properties of this neuron are, thus, easily explained by assuming that one of the presynaptic neurons is an FM-specialized neuron selectively sensitive to a downward-sweeping FM sound. Other combinations of signals (e.g., H_1–H_1, H_2–H_2) had no effect on this neuron. For the maximum excitation of this neuron, CF_1 should be 27.80 kHz and FM_2 should sweep from 61.74 to 49.74 kHz (Fig. 6.16C) and their amplitude should be 71 and 43 dB SPL, respectively (Fig. 6.16D). The presynaptic neurons apparently show a nonmonotonic impulse-count function as a result of lateral inhibition. The facilitation-frequency-tuning curves for CF_2 and CF_3 of Fig. 6.15 are very sharp. The slopes of the curves are nearly vertical and their bandwidth is narrow even at 40–70 dB above the minimum threshold. Such sharp curves are easily explained by assuming that the excitatory areas of presynaptic neurons are sharpened by lateral in-

---

FIG. 6.16. (continued)

FM_1. The $FM_2 \downarrow$ and $FM_2 \uparrow$ are 56 dB SPL and sweep in the range 49.5–61.5 kHz. The NB is 56 dB SPL and 49.5–61.5 kHz bandwidth. Each PST histogram consists of neural activity for 100 presentations of an identical sound or sounds. B, Cumulative histograms of the responses and background discharges, comparable to these shown in A. Each histogram is the average of two samples of 100 presentations. Ca and Cb, respectively, represent the CF_1 and FM_2 facilitation areas (facilitation-frequency-tuning curves). In order to measure the range of the 30-ms-long CF_1 to be combined with the FM_2 for the facilitation of the neuron, the latter was fixed near the parameters for the best facilitation and the amplitude range of the former was measured as a function of frequency. The threshold of the response was defined to be the stimulus amplitude, which evoked 0.1 impulse/stimulus on the average. The range of the 4-ms-long FM_2 to be combined with the CF_1 for facilitation was measured by varying the amplitude and initial and terminal frequencies of a 12 kHz frequency sweep (center frequency of sweep is plotted). The parameters of the CF_1 were set near those for the best facilitation. This neuron had no excitatory area. D, Impulse-count function measured as a function of either the CF_1 (a) or FM_2 (b). One of them was fixed in amplitude as in C and the number of impulses per stimulus was measured as a function of the amplitude of the other. The data were obtained from a lightly anesthetized animal between 3 and 6 h after Nembutal administration (Suga, O'Neill and Manabe, 1978, ref. 41; by permission of the Amer. Assoc. Adv. Sci.).

hibition. Some presynaptic neurons are probably qualified to be called CF-specialized neurons.

It is a fact that *the auditory system of the mustached bat contains both circuits which extract information-bearing elements and those which assemble them.* It is not difficult to hypothesize neural network models which incorporate these two sets of information, but anatomical data in support of these models are not yet obtained.

In spoken Engish, various combinations of formants and/or transitions carry different information. Formants and transitions may be called CF and FM components, respectively. Since FM_1–FM_n and CF_2–FM_2 facilitation neurons are specialized to examine particular sequences of signal elements and CF_1/CF_n facilitation neurons are specialized to examine overtone structure of a signal (19, 39, 41, 42), we may hypothesize that spoken English is eventually processed by neurons sensitive to particular combinations of information-bearing elements. At present, there is no direct technique to test this hypothesis. Therefore, the validity of this hypothesis can only be understood in terms of comparative auditory physiology.

A male bullfrog, *Rana catesbeiana*, produces croaking calls, each characterized by two major components at 200 and 1400 Hz. At the periphery, there are two groups of neurons, each tuned to one of these components; their simultaneous excitation is essential to evoke vocal response from the male (1, 6). Neurons in the dorsal thalamus show clear facilitation of responses by simultaneous presentation of the two major components in the call (17). This is particularly interesting, because even in the frog the auditory system has neurons sensitive to combinations of information-bearing elements, even though peripheral neurons show phase-locked discharges to low-frequency acoustic signals.

The auditory system of each species is (including man) specialized for analyzing and processing the sounds which are important to it, so that some of the data obtained from one species may be unique and may be hardly applicable to other species; however, other data, related to the most fundamental neural mechanisms, may be shared with other species. Neural mechanisms for processing complex sounds by examining combinations of the information-bearing elements may be common among the central auditory systems of different species of higher vertebrates. For better understanding of neural mechanisms for processing auditory information, comparative studies with different species of animals should be performed on (i)physical properties and biological significance of sounds used by a species, (ii) information-bearing parameters (IBPs) or elements in them, (iii) synchronous discharges of cortical neurons with these sounds or information-bearing elements, (iv) neurons tuned to particular values of IBPs, (v) neurons specialized for responding to particular combinations of IBPs and (vi) the functional organization of the auditory system.

For auditory physiology, those animals for whom hearing is especially important are very valuable, because the neural mechanisms involved in the specialization of the auditory system for processing particular types of information can be explored much more easily. Among the recent discoveries made in the mustached bat and barn owl, the following six are particularly significant: (A) the peripheral auditory system can have a series of sharply frequency-tuned filters with slopes of 1500 to 1800 dB/octave and quality factor of 210 (37, 47); (B) the central auditory system can have neural circuits producing neurons tuned to particular values of IBPs (30); (C) the central auditory system can have neural circuits producing neurons sensitive to particular combinations of IBPs (19, 41, 42); (D) neurons sensitive to different combinations of IBPs are in different clusters located at indentifiable loci in the brain (39, 41, 42); (E) the central auditory system can have axes or coordinates for systematical representation of IBPs such as stimulus frequency (36), stimulus amplitude (31), time interval between signals (39), frequency relationship between signals (54) and direction of a sound source (14) by the location of activated neurons; and (F) each of such axes—population of neurons—representing an IBP is apportioned according to the biological importance of the IBP. The amplitopic and odotopic representations found in the mustached bat (31, 39) and neural map of the auditory space found in the barn owl (14) are the result of interactions of excitatory and inhibitory neurons occuring in the spatial and temporal domains. Inhibition can theoretically improve the neural representation of auditory information by introducing contrast in the spatiotemporal pattern of neural activity.

3.3. Specialized and Unspecialized Neurons

In several different species of animals in addition to bats, it has been demonstrated that, unlike the periphery, the central auditory system contains neurons that are apparently specialized to respond only to biologically significant signals in a certain narrow range of particular parameters in the time, frequency and amplitude domains, or the interaural disparity between these domains. An important question, then, is whether the activity of such specialized neurons can be correlated with categorical perception or the detection of individuality of a signal. The answer to this question depends upon whether *the bandwidth of the neural filter* is comparable to *the bandwidth of a behavioral filter* for categorical perception or the detection of individuality. In most cases, the properties of specialized neurons have not been adequately described, because of the absence of studies not only on such aspects as their *level tolerance* and *rejection mode*, but also on *filter bandwidth* (32). Furthermore, the behavioral filter bandwidth for categorical or individuality perception of biologically significant signals has only partly been studied in animals. Thus, a comparison of a neural filter with a behavioral filter is not possible at present. Accordingly, it is impossible to discuss critically whether or not specialized neurons found thus far are directly related to categorical perception or detection of individuality of biologically significant signals.

In this article, the discussion is focused on specialized neurons, because clues for understanding neural mechanisms for information processing are hidden in the differences in response properties of neurons among and within individual auditory nuclei. When we focus on one aspect of the auditory system, we do not necessarily describe other aspects. Therefore, I should like to stress that in individual auditory nuclei, including the auditory cortex, not only specialized but also unspecialized neurons exist, and that some unspecialized neurons in the auditory cortex are somewhat comparable to primary auditory neurons in their response properties.

Why are there "primary-like" auditory neurons, in addition to specialized neurons, even in the auditory cortex? I speculate that the advantage of specialized neurons lies in their ability to perform quick informaion processing, while avoiding excitation by biologically less-significant sounds. If every neuron were specialized, the auditory system would lose all information except for that extracted by specialized neurons. It may be desirable for the auditory cortex to contain some primary-like auditory neurons in order to maintain signals coded by primary auditory neurons as intact as possible. Decoding by the spatiotemporal pattern of activity of these neurons may be slower than that by specialized neurons, but this mode of decoding may be essential for processing acoustic signals that are less familiar or unfamiliar to the animal (32).

3.4. "Parallel-Hierarchical" Processing

In mammals, an auditory signal sent to the brain by the cochlear nerve ascends from the cochlear nucleus to the auditory cortex of the cerebrum through many intermediate nuclei. The auditory signal is projected in parallel to each level of the central auditory system, because of multiple projections from the cochlea. These multiple levels and multiple projections indicate that different types of auditory information or different attributes of the signal are processed both hierarchically and in parallel. This should be true in the auditory system regardless of whether it is concerned with communication or echolocation. For echolocation, the auditory system should process several different types of biosonar information related to the different attributes of a target.

As summarized in this article, different types of biosonar information are represented in different areas of the cerebral cortex. For amplitopic representation, it is absolutely necessary to have neurons that show nonmonotonic impulse-count functions. Such neurons have been found in the inferior colliculus (8, 26). For odotopic representation, the brain should have neurons that show facilitation of responses to echoes returning after their recovery periods, and for those that respond only to echoes following orientation sounds with particular delays. Such neurons have been found in the midbrain (2, 9, 44). Therefore, amplitopic and odotopic representations may exist in subcortical auditory nuclei to some extent. Harmonic-sensitive neurons may also be found in subcortical auditory nuclei in the future. It is, however, very likely that more specialized neurons are found more frequently at higher levels of the auditory system (30). Many neurons in the FM–FM area, for instance, are sharply tuned to a particular echo delay and a particular echo amplitude. Doppler shift is also an influential parameter for their excitation. Such response properties are not readily explained only by interaction among primary and primary-like auditory neurons. It is very unlikely that such neurons exist in the cochlear nuclei.* But their response properties probably result from neural interaction taking place at higher auditory nuclei. Therefore, the neurophysiological data also indicate that different types of biosonar information are processed both hierarchically and in parallel. *Processing of auditory information is probably neither only parallel nor only hierarchical, but "parallel-hierarchical".*

"Simple" hierarchical organization implies (i) the progressive increase in specialization of neurons at higher levels of auditory nuclei, and probably automatically implies (ii) the presence of "detectors" specific to individual biologically significant sounds at the summit of the hierarchy. Hierarchical organization is, therefore, one of the possible neural mechanisms for production of detectors required by the detector hypothesis. The concept of hierarchical or-

---

*In lower auditory nuclei, such as the dorsal cochlear nucleus, specialized neurons similar to those at higher levels may be found in the future. Such a finding may, however, not be necessarily contrary to this notion, because the nuclei receive efferent nerve fibers originating from the higher auditory nuclei.

ganization is "convenient" for explaining the response properties of specialized neurons at higher levels and is valuable for construction of one of the neural network models for signal processing.

Since the auditory system is anatomically complex, some parts of the electrophysiological data obtained from it are easily explained without the concept of hierarchical organization, while other parts are difficult to explain without it. It is, however, a problem whether specialized neurons found in the auditory cortex (e.g., range-sensitive neurons) can be interpreted as detectors in the hierarchical organization. As described in the next section, it is more appropriate to interpret them as filters acting as a kind of cross-correlator (34).

3.5. Specialized Neurons as IBP Filters

In the primary auditory cortex of cats and monkeys, some neurons are apparently tuned not only in frequency, but also in amplitude. Amplitopic representation, however, has not yet been studied.* If the frequency vs amplitude coordinates are not found in both the areas anterior and posterior to the DSCF area (Fig. 6.5) and also in the auditory cortex of other animals (e.g., cats and monkeys) that use broad-band acoustic signals, we must consider that the coordinates found in the DSCF area are unique in the mustached bat, because of its use of echolocation, and that the data do not necessarily support the amplitude-spectrum hypothesis, but favor the IBP filter hypothesis, because neurons in the DSCF area are specialized to respond to particular combinations of two acoustic parameters that are related to either subtended target angle or target velocity information. If the coordinates are found in the other mammals, on the other hand, we must consider that the data obtained from the DSCF area basically fit the amplitude-spectrum hypothesis, but as an exceptional case the data favor the IBP filter hypothesis, because the activity of individual neurons in the coordinates represents certain information important for echolocation.

As described above, neurons in the DSCF area are specialized to respond to *particular combinations* of two acoustic parameters and are "simple" compared with those in the FM–FM and CF/CF areas, but share common properties to some extent, because the

*By evoked potential studies, Tunturi (48) discovered that the dog's auditory cortex had two coordinates, frequency and threshold. Curiously, this threshold representation is only applied to the input from the ipsilateral ear, which is not the main excitatory input of the cortex. If cortical neurons commonly showed monotonic impulse-count functions, this threshold representation would be a kind of amplitopic representation. This, however, appears not to be the case. Further studies remain to be performed (see Section 1-1).

latter are specialized to respond to particular combinations of two signal components. Thus, neurons in all these three areas appear to favor the detector hypothesis at first glance. What our data thus far obtained actually support, however, is not precisely the detector hypothesis but a compromise that incorporates both the amplitude-spectrum and detector hypotheses and that is called the *IBP filter hypothesis* (Fig. 6.1., part 3).

To generalize what has recently been found in the mustached bat (19, 31, 36, 39, 41, 42) and barn owl (14), the term *information-bearing parameter (IBP)* is introduced, because the term *information-bearing element* does not include interaural time and amplitude differences, interval between signals (e.g., echo delay) and other parameters characterizing combinations of information-bearing elements that are important for communication and/or echolocation. *An IBP is the limited part of a continuum that carries information important for the species in nature.* An identical IBP can be quite different in its biological significance for different species of animals.

The IBP filter hypothesis states that *specialized neurons expressing the outputs of neural circuits tuned to particular IBPs or particular combinations of IBPs are understood to be IBP filters* (a kind of cross-correlator) and that *information processing is performed by many IBP filters that are systematically arranged in a hypothetical center* according to their filter properties (32, 34). The data obtained from the auditory cortex of the mustached bat can easily be generalized according to this hypothesis and can complement the hypothesis as follows. (i) The central auditory system contains specialized neurons (IBP filters), which are tuned to particular IBPs or combinations of IBPs, and which pass signals with particular IBPs or particular combinations of IBPs. (ii) Different types of IBP filters are found in separate clusters depending upon their function. (iii) Within individual clusters, IBP filters differ from one another in tuning properties and can be arranged to represent IBPs systematically. Thus, they can form an axis or coordinate system within individual clusters. (iv) More significant parts of an IBP are overrepresented by large numbers of IBP filters within a cluster. In other words, the more important part is represented with higher resolution. (v) Interaction between excitation and inhibition occurring spatially and temporally plays a key role in producing IBP filters and increases the sharpness of the IBP filters.* Consequently, the contrast in neural representation of audi-

---

*Lateral inhibition is usually thought of in the spatial domain. To understand neural mechanisms for some of the data obtained from the bat, it should also be considered in the time domain.

tory information in a cluster increases. (vi) Any filter has a particular bandwidth. Widths of IBP filters are not so narrow that only a few neurons or neurons in a single cortical column are selectively excited by a particular signal. Therefore, many IBP filters in several cortical columns are activated by the signal. Furthermore, individual IBP filters thus far studied may not be sharp enough to explain the signal recognition or discrimination capability of an animal, so that the spatiotemporal pattern of activity of IBP filters that appear along an IBP axis or in IBP coordinates is probably directly related to signal recognition. Minor differences in the spatiotemporal pattern of activity are probably related to recognition of individuality of a signal.

For pattern recognition, information about the acoustic pattern to be recognized should be stored in the brain and should be compared with incoming signals. That is, the acoustic pattern should be theoretically cross-correlated with stored information. (vii) Since a filter acts as a kind of cross-correlator, we may consider that IBP filters are cross-correlators that correlate acoustic signals with their filter properties, i.e., stored information, and that the degree of cross-correlation is expressed by the magnitude of the output of the IBP filters. In other words, neurons show the maximum excitation only when the properties of acoustic signals perfectly match their filter properties.

Considering the above, the functional organization of the DSCF, FM–FM and CF/CF areas may be summarized as follows: (A) In the DSCF area, IBP filters (correlators) are systematically arranged in the coordinates representing target velocity and subtended target angle information. The IBP filters in the ventral and dorsal parts are, respectively, suited for target detection and localization. The IBP filters in the anterior and posterior halves are speculated to be suited for processing either fluttering or nonfluttering targets and to form two functional subdivisions. (B) In the FM–FM area, on the other hand, IBP filters tuned to particular combinations of different FM components are found in three different aggregates and, in each aggregate, they are systematically arranged along an axis representing target range. We do not yet know if another auditory parameter is systematically arranged in an axis normal to the range axis, but we may speculate that target characteristics are represented by distribution of neural activity along it (Fig. 6.5). FM_1–FM_4 neurons are theoretically better suited for fine characterization of small targets than FM_1–FM_2 neurons. (C) In the CF/CF area, IBP filters tuned to combinations of different CF components are found in two different clusters. In each cluster, there are coordinates representing the frequencies of the two CF components necessary for facilitation (Fig. 6.5). We may speculate that a Doppler-shift, i.e., target velocity in the radial direction is systematically represented in these coordinates (53,54).*

Different subdivisions of the auditory cortex are interconnected with association fibers, and the auditory cortices in both hemispheres with commissural ones (Fig. 6.4B). The fundamental role of these fibers has not yet been experimentally explored in any animal. What is summarized in this article is the activity of the neurons that receive thalamic afferent fibers and interact with many other cortical neurons through association and commissural fibers. (Fig.6.4B). How is the neural activity of cortical auditory neurons related to recognition of an overall acoustic image? One of our working hypotheses is that it is directly related to neural activity in an area integrating all subdivisions specialized for processing different types of auditory information, while an alternative hypothesis is that it is directly related to the spatiotemporal pattern of neural activity occurring at these functional subdivisions. The latter appears to be more likely. At the present stage of our research, however, we do not know the upper limit of neural specialization in response to complex acoustic signals. We obviously must further pursue the problems of the neural representation of auditory information by exploring the origin and destination of information through neurons in each subdivision.

3.6. Appendix

It should be noted that the IBP filter hypothesis incorporates both the amplitude–spectrum and detector hypotheses, but this does not exclude the synchronization hypothesis. The synchronization hypothesis is probably valid to explain certain aspects of sound recognition. For instance, perception of beating insect wings may be based upon discharges of neurons synchronized with the wing beat, and pitch perception is presumably based upon discharges of neurons synchronized with fundamentals of vowels and other complex sounds. In general, perception related to low repetition rates of sounds (sound waves) is probably directly mediated by stimulus or phase-locked discharges. The functional organization of the

---

*It should be noted that the fascinating data obtained from the mesencephalic laterodorsal nucleus of the barn owl *Tyto alba* (14) also fits into the IBP filter hypothesis, because (i) neurons tuned to particular azimuthal and elevational degrees examine combinations of interaural amplitude and time differences, (ii) they aggregate at the anterolateral part of the nucleus, (iii) they are arranged along the axes for the systematic representation of the auditory space in front of the animal, and (iv) this neural map of the auditory space overrepresents a certain portion of the auditory space that is important for the animal.

FM–FM area, however, suggests that even the temporal pattern of discharge can be eventually expressed by the loci in the brain, when it carries essential information for an animal. Stimulus-locked discharges appear to be most prominent at the periphery. With ascent of impulses along the auditory system, it may become generally poor because of synaptic transmission. For exploring the auditory neural mechanisms, therefore, it is essential to study response properties of higher-order neurons beyond stimulus-locking. In such studies, biologically significant sounds should be used as stimuli.

4. Summary and Conclusions

1. In the cerebral cortex of the mustached bat, *Pteronotus parnellii rubiginosus*, there are multiple projections of the cochlea with and without overlap. In the areas where different parts of the cochlea project with overlap, the tonotopic representation is vague and neurons are sensitive to particular combinations of elements in complex sounds. To explore the functional organization of such an auditory cortex, experiments were performed with lightly anesthetized and with unanesthetized mustached bats. The data obtained under these two conditions were very similar. In terms of the responses of single neurons to constant frequency (pure) tones and frequency-modulated sounds, three functionally distinct areas have been found: DSCF, FM–FM and CF/CF areas (Fig. 6.4). Since the neuroethological approach has proved successful in exploring the functional organization of sensory systems, stimuli were delivered mimicking orientation sounds (biosonar signals) and echoes. Each of these sounds consists of up to four harmonics, comprised of constant-frequency (CF) and frequency-modulated (FM) components (Fig. 6.2A).

2. The DSCF area (about 30% of the primary auditory cortex) is devoted to processing information carried by the main CF component (61–63 kHz) of Doppler-shifted echoes. It has a radial frequency axis representing target–velocity information and a circular amplitude axis representing subtended target angle information. Along the frequency axis, best frequency changes at a rate of 20–30 Hz/neuron, i.e., velocity information is represented by increments of 5.6–8.4 cm/s/neuron. Most neurons in this area are extremely sensitive to the frequency modulation of an echo that may be produced by a flying insect, and show discharges synchronized with the modulation as small as ±0.01%, i.e., ±6 Hz shift for 61 kHz. The systematic representation of stimulus amplitude by

the location of activated neurons is called *amplitopic representation*. The dynamic range of the amplitopic representation is about 70 dB. In terms of binaural interaction, the DSCF area consists of two functional subdivisions that are occupied mainly by either E-E or I-E neurons and are accordingly suited for either target detection or localization (Figs. 6.5–6.9).

3. In the FM–FM area, neurons are specialized for responding to particular combinations of FM components in orientation sounds and echoes. For their excitation, an important stimulus parameter is echo delay, which is the primary cue for target ranging. Therefore, these neurons are functionally sensitive to target range. Two classes of *range-sensitive neurons* were found: (i) *tracking neurons*, whose preferred echo delay *(best delay)*, i.e., preferred target range *(best range)*, for response to an echo following the orientation sound becomes shorter and narrower as the bat closes in on the target and (ii) *range-tuned neurons*, whose best range is constant, responding to the target only when it is within a certain narrow fixed range. Range-tuned neurons are specialized for processing echoes from targets at particular ranges and are systematically arranged according to their best ranges, so that they form a neural axis representing target range from 7 to 310 cm. This is called *odotopic representation.* Best range varies at a rate of 2.0 cm/neuron along the range axis. Many range-tuned neurons are tuned not only to a specific echo delay, but also to a particular echo amplitude. That is, they are specialized to respond best to targets with particular cross-sectional areas at particular distances. Furthermore, they respond better to Doppler-shifted echoes from approaching targets. The auditory cortex is undoubtedly involved in information processing even in the terminal phase of echolocation (Figs. 6.5 and 6.10–6.13).

4. In the CF/CF area, neurons are specialized for responding to particular combinations of CF components that are harmonically or quasiharmonically related. The amplitude relationship between the CF components to be combined is very important in some neurons for their excitation. Since the neurons are broadly tuned to echo delay of 0–10 ms, their putative function is to detect target velocity, not target range (Figs.. 6.5, 6.14 and 6.15).

5. Neurons in both the FM–FM and CF/CF areas are specialized to "examine" or "assemble" combinations of information-bearing elements. Neurons examining different combinations are found in different aggregates. Each aggregate in the FM–FM area has a range axis, while that in the CF/CF area has frequency vs frequency axes (Fig. 6.5). The axis or axes are apportioned according to the biological importance of an information-bearing parameter. It should be stressed that in these two areas, tonotopic and amplitopic repre-

sentations are vague, that responses of neurons to single pure tones and FM sounds are poor, and that the functional organization of these areas would not have been discovered, had not biologically significant sounds been used as stimuli.

6. There is a remarkable contrast in functional organization between the DSCF and the *complex-sound processing* (FM–FM and CF/CF) *areas.* The amplitude spectrum of acoustic signals is represented by the spatiotemporal pattern of activity of neurons arranged in the coordinates of the DSCF area as predicted by the *amplitude-spectrum hypothesis,* while combinations of important signal elements are represented by that of neurons in the complex-sound processing areas, as somewhat similar to the prediction of the detector hypothesis (Fig. 6.1). An appropriate interpretation of the functional role of neurons in all these three areas is, however, that they are *IBP (information-bearing parameter) filters* tuned to particular IBPs or combinations of IBPs, and that they act as a kind of cross-correlator that correlates incoming signals with their filter properties. Our data best fit the *IBP filter hypothesis.*

7. The auditory system contains both circuits that extract information-bearing elements and those that assemble or examine them. It is complex anatomically and physiologically and processes auditory information both hierarchically and in parallel. Four working hypotheses concerning the neural basis of acoustic pattern recognition (amplitude spectrum, detector, IBP-filter and synchronization hypotheses) are probably not exclusive, but individually valid depending upon the type of auditory information and species.

Acknowledgments

I am grateful for the support of NSF (grant BNS78-12987) and for the cooperation of E. G. Jones, T. Manabe, W. E. O'Neill, J. Ostwald and P. Wasserbach. I am also thankful to C. N. Woolsey for his effort in editing this book and R. W. Coles for reading this article.

References

1. CAPRANICA, R. R. Vocal responses of the bullfrog to natural and synthetic mating calls. *J. Acoust. Soc. Amer.,* 40: 1131–1139, 1966.
2. FENG, A. S., SIMMONS, J. A., AND KICK, S. A. Echo detection and target-ranging neurons in the auditory system of the bat *Eptesicus fuscus. Science,* 202: 645–648, 1978.

3. FENTON, M. B. Variation in the social calls of little brown bats *(Myotis lucificus). Canadian J. Zool.,* 55: 1151–1157, 1977.

4. FENTON, M. B. Adaptiveness and ecology of echolocation in terrestrial (aerial) systems. In: *Biosonar Systems,* edited by R. G. BUSNEL, AND J. F. FISH, New York: Plenum, 1980, pp. 427–446.

5. FRIEND, J. H. SUGA, N., AND SUTHERS, R. A. Neural responses in the inferior colliculus of echolocating bats to artificial orientation sounds and echoes. *J. Cell Physiol.,* 67: 319–332, 1966.

6. FRISHKOPF, L. S., AND GOLDSTEIN, M. H., JR. Responses to acoustic stimuli from single units in the eighth nerve of the bullfrog. *J. Acoust. Soc. Amer.,* 35: 1219–1228, 1963.

7. GOLDMAN, L. J., AND HENSON, O. W., JR. Prey recognition and selection by the constant frequency bat, *Pteronotus p. parnellii. Behav. Ecol. Sociobiol.,* 2: 411–419, 1977.

8. GRINNELL, A. D. The neurophysiology of audition in bats: intensity and frequency parameters. *J. Physiol., London,* 167: 38–66, 1963.

9. GRINNELL, A. D. The neurophysiology of audition in bats: temporal parameters. *J. Physiol., London,* 167: 67–96, 1963.

10. GRINNELL, A. D. Comparative auditory neurophysiology of neotropical bats employing different echolocation signals. *Z. vergl. Physiol.,* 68: 117–153, 1970.

11. GRINNELL, A. D. AND GRIFFEN, D. R. The sensitivity of echolocation in bats. *Biol. Bull., Woods Hole.,* 114: 10–22, 1958.

12. HENSON, O. W. JR. The activity and function of the middle ear muscles in echolocating bats. *J. Physiol., London,* 180: 871–887, 1965.

13. KATSUKI, Y., WATANABE, T., AND SUGA, N. Interaction of auditory neurons in response to two sound stimuli in cat. *J. Neurophysiol.,* 22: 603–623, 1959.

14. KNUDSEN, E. I., AND KONISHI, M. Space and frequency are represented separately in auditory midbrain of the owl. *J. Neurophysiol.,* 41: 870–884, 1978.

15. MANABE, T., SUGA, N., AND OSTWALD, J. Aural representation in the Doppler-shifted CF processing area of the primary auditory cortex of the mustache bat. *Science,* 200: 339–342, 1978.

16. MOUNTCASTLE, V. B. Modality and topographic properties of single neurons of cat's somatic sensory cortex. *J. Neurophysiol.,* 20: 408–434, 1957.

17. MUDRY, K. M., CONSTANTIN-PATON, M., AND CAPRANICA, R. R. Auditory sensitivity of the diencephalon of the leopard frog *Rana p. pipiens. J. Comp. Physiol.,* 114: 1–13, 1977.

18. NOVICK, A., AND VAISNYS, J. R. Echolocation of flying insects by the bat, *Chilonycteris parnellii. Biol. Bull.,* 127: 478–488, 1964.

19. O'NEILL, W. E., AND SUGA, N. Target range-sensitive neurons in the auditory cortex of the mustache bat. *Science,* 203: 69–73, 1979.

20. SCHNITZLER, H.-U. Echoortung bei der Fledermaus *Chilonycteris rubiginosa. Z. vergl. Physiol.,* 68: 25–38, 1970.

21. SCHULLER, G. The role of overlap of echo with outgoing echolocating sound in the bat *Rhinolophus ferrumequinum. Naturwissenchaften.*, 61: 171–172, 1974.

22. SIMMONS, J. A. The sonar receiver of the bat. *Ann. NY Acad. Sci.*, 188: 161–174, 1971.

23. SIMMONS, J. A. Perception of echo phase information in bat sonar. *Science*, 204: 1336–1338, 1979.

24. SIMMONS, J. A., HOWELL, D. J. AND SUGA, N. The information content of bat sonar echoes. *Amer. Scient.*, 63: 204–215, 1975.

25. SIMMONS, J. A., AND O'FARRELL, M. J. Echolocation by the long-eared bat, *Plecotus phyllotis. J. Comp. Physiol.*, 122: 201–214, 1977.

26. SUGA, N. Analysis of frequency modulated sounds by auditory neurones of echolocating bats. *J. Physiol., London*, 179: 26–53, 1965.

27. SUGA, N. Functional properties of auditory neurones in the cortex of echolocating bats. *J. Physiol., London*, 181: 671–700, 1965.

28. SUGA, N. Analysis of frequency-modulated and complex sounds by single auditory neurones of bats. *J. Physiol., London*, 198: 51–80, 1968.

29. SUGA, N. Classification of inferior collicular neurones of bats in terms of responses to pure tones, FM sounds, and noise bursts. *J. Physiol., London*, 200: 555–574, 1969.

30. SUGA, N. Feature extraction in the auditory system of bats. In: *Basic Mechanisms in Hearing*, edited by A. R. MØLLER. *New York:* Acad. Press, 1973, pp. 675–744.

31. SUGA, N. Amplitude-spectrum representation in the Doppler-shifted-CF processing area of the auditory cortex of the mustache bat. *Science*, 196: 64–67, 1977.

32. SUGA, N. Specialization of the auditory system for reception and processing species-specific sounds. *Fed. Proc.*, 37: 2342–2354, 1978.

33. SUGA, N. Representation of auditory information by the brain(I). *Shizen, Chuokoron-sha, Tokyo, Japan*, 79-5: 26–41, 1979 (in Japanese).

34. SUGA, N. Representation of auditory information by the brain(II). *Shizen, Chuokoron-sha, Tokyo, Japan*, 79-6: 70–81, 1979 (in Japanese).

35. SUGA, N., AND JEN, P. H.-S. Peripheral control of acoustic signals in the auditory system of echolocating bats. *J. Exptl. Biol.*, 62: 277–311, 1975.

36. SUGA, N., AND JEN, P. H.-S. Disproportionate tonotopic representation for processing species-specific CF-FM sonar signals in the mustache bat auditory cortex. *Science*, 194: 542–544, 1976.

37. SUGA, N., AND JEN, P. H.-S. Further studies on the peripheral auditory system of "CF-FM" bats specialized for the fine frequency analysis of Doppler-shifted echoes. *J. Exptl. Biol.*, 69: 207–232, 1977.

38. SUGA, N., AND MANABE, T. Neural basis of amplitude-spectrum representation in the auditory cortex of the mustached bat. *J. Neurophysiol.* (in press).

39. SUGA, N., AND O'NEILL, W. E. Neural axis representing target range in the auditory cortex of the mustached bat. *Science*, 206: 351–353, 1979.

40. SUGA, N., AND O'NEILL, W. E. Auditory processing of echoes: representation of acoustic information about the environment in the brain of a bat. In: *Biosonar Systems*, edited by R. G. BUSNEL AND J. F. FISH. New York: Plenum, 589–611, 1980.

41. SUGA, N., O'NEILL, W. E., AND MANABE, T. Cortical neurons sensitive to particular combinations of information bearing elements of biosonar signals in the mustached bat. *Science*, 200: 778–781, 1978.

42. SUGA, N., O'NEILL, W. E., AND MANABE, T. Harmonic-sensitive neurons in the auditory cortex of the mustached bat. *Science*, 203: 270–274, 1979.

43. SUGA, N., AND SCHLEGEL, P. Neural attenuation of responses to emitted sounds in echolocating bats. *Science*, 177: 82–84, 1972.

44. SUGA, N., AND SCHLEGEL, P. Coding and processing in the auditory systems of FM-signal-producing bats. *J. Acoust. Soc. Amer.*, 54: 174–190, 1973.

45. SUGA, N., SCHLEGEL, P., SCHIMOZAWA, T., AND SIMMONS, J. A. Orientation sounds evoked from echolocating bats by electrical stimulation of the brain. *J. Acoust. Soc. Amer.*, 54: 793–797, 1973.

46. SUGA, N., AND SHIMOZAWA, T. Site of neural attenuation of responses to self-vocalized sounds in echolocating bats. *Science*, 183: 1211–1213, 1974.

47. SUGA, N., SIMMONS, J. A., AND JEN, P. H.-S. Peripheral specialization for fine analysis of Doppler-shifted echoes in "CF-FM" bat *Pteronotus parnellii. J. Exptl. Biol.*, 63: 161–192, 1975.

48. TUNTURI, A. R. A difference in the representation of auditory signals for the left and right ears in the iso-frequency contours of the right middle ectosylvian auditory cortex of the dog. *Amer. J. Physiol.*, 168: 712–727, 1952.

49. WOOLSEY, C. N., AND WALZL, E. M. Topical projection of nerve fibers from local regions of the cochlea to the cerebral cortex of the cat. *Amer. J. Physiol.*, 133, 498–499, 1941.

50. BRUGGE, J. F., AND MERZENICH, M. M. Patterns of activity of the auditory cortex. In *Basic Mechanisms in Hearing*, edited by A. R. MØLLER. *New York*: Academic Press, 1973. pp. 745–766.

51. MERZENICH, M. M., KNIGHT, P. L., AND ROTH, G. L. Representation of cochlea within primary auditory cortex in the cat. *J. Neurophysiol.*, 38: 231–249, 1975.

52. ROSE, J. E., GROSS, N., GEISLER, C. D., AND HIND, J. E. Some neural mechanisms in the inferior colliculus of the cat which may be rele-

vant to localization of a sound source. *J. Neurophysiol.*, 29: 288–314, 1966.

53. SUGA, N. Cortical representation of biosonar information in the mustached bat. In *Sensory Function, Adv. Physiol. Sci.* vol. 16, edited by E. Grastyan and P. Molnar, New York: Pergamon, 1981, 119–125.

54. SUGA, N., KUJIRARI, K., AND O'NEILL, W. E. How biosonar information is represented by the bat's cerebrum. In *Neuronal Mechanisms of Hearing*, edited by J. Syka and L. Aitkin. New York: Plenum, 197–219. 1981.

55. WEVER, E. G., *Theory of Hearing.* New York: Wiley, 1949.

Chapter 7

A Theory of Neural Auditory Space

Auditory Representation in the Owl and Its Significance

*Masakazu Konishi and Eric I. Knudsen**

*Division of Biology, California Institute of Technology
Pasadena, California*

1. Introduction

The topographical projection of sensory epithelia onto central neural structures is one of the most striking design features of the brain. This type of neuronal connectivity gives rise to the "maps" of the retina, cochlea and body surface in the central nervous system. Numerous studies have addressed themselves to the developmental

*Present address: Department of Neurobiology, Stanford University School of Medicine, Stanford, California 94305.

219

mechanisms underlying the formation of topographical projections. Partly because of these trends, there has been relatively little
discussioon of mapping as a physiological phenomenon rather
than as an anatomical one. To the extent that the organization of a
sensory representation strictly reflects the brain's architectural requirements, including such factors as simple physical packing and
routing and various embryological rules, questions about mapping
reduce to those about the underlying anatomical projection; in this
case, a physiological map is a byproduct of the brain's architectural
design.

An alternative to this argument is that the construction of a
physiological map may be the primary aim of at least some topographical projections. In this case, the systematic mapping of a
neuronal response property itself has physiological significance.
However, it is not possible to discern which argument applies as
long as physiological mapping and anatomical projections are
equivalent. Therefore, the demonstration of a physiological map
that is not derived from topographically ordered anatomical sub-
strates would constitute an important step towards the recognition
of mapping as a physiological phenomenon.

The neural map of auditory space in the midbrain of the barn
owl (*Tyto alba*) is such a map. Sound localization by the owl involves the computation of sound locations from binaural inputs.
The computed locations are "topographically projected" onto the
owl's brain. The main aim of this article is to discuss theoretical implications of the auditory space map. Other aspects of this study,
such as the mechanism of sound localization and methodological
details, are included insofar as they are necessary to describe the
characteristics and significance of the map.

2. Perception of Auditory Space

How the owl perceives auditory space can be assessed by studying
its sound localization behavior. The barn owl has a natural tendency to turn its head toward a sound source, which provides a reliable measure of its ability to localize sound (6). The owl evidently
uses a binaural method to localize sound, for monaural occlusion
causes systematic errors in sound localization; when its right ear is
plugged, the owl errs by orienting to the left and below the target,
and when its left ear is plugged, to the right and above it. The owl
determines the elevation of sound by using binaural intensity difference and its azimuth by binaural time difference (7, 8).

The owl needs only one set of binaural disparities to determine
the location of a sound source. This is demonstrated by the owl's

ability to localize a target that ceases to emit sound prior to the onset of head-turning. This paradigm allows the owl to measure binaural disparities only at one head position. The owl must have a scheme to translate each set of binaural disparities into a location in space. The scheme must contain a predetermined set of disparity values assigned to each location. Thus, auditory space, as perceived by the owl using both ears, should consist of a matrix of binaural disparities. The accuracy of determining a location decreases systematically as azimuthal and elevational positions depart from the point directly in front of the face (Fig. 7.1). The owl presumably makes errors because of uncertainty in the exact values of the binaural disparities. This suggests that the spatial density of discriminable disparities, hence discriminable locations, decreases systematically as a function of angular distance from the midpoint of the face.

3. Auditory Receptive Field

"The receptive field of a cell in the visual system may be defined as the region of retina (or visual field), over which one can influence the firing of that cell" (1). In the visual system, an area of space casts its image on the retina. Since this relationship between the sensory surface and extrapersonal space does not exist in the auditory system, it might seem unlikely to find auditory neurons having receptive fields.

However, there is convincing evidence that the receptive field concept is applicable to auditory space. For example, the visual receptive fields of cat auditory–visual bimodal neurons obviously satisfy the definition of a receptive field. Since the same neuron responds to auditory stimuli located only in or near the area of space coincident with its visual field, it would seem reasonable to call this area the "auditory receptive field" of that neuron (11).

The above observation suggests that the owl's auditory space might be represented by the receptive fields of single neurons. A direct test of this expectation is to study the response of auditory neurons to changes in the location of a sound source. Departing from the traditional method of presenting binaural stimuli via ear tubes, we use a device to move a small speaker around the owl's head in both azimuth and elevation under free field conditions.

A unit's receptive field is defined as the area of space in which appropriate sound stimuli excite the unit and is determined in the following manner: when an auditory unit is isolated, the speaker is moved in space to find the location where the unit's discharge rate is maximal. The thresholds of the unit to noise and tone bursts are

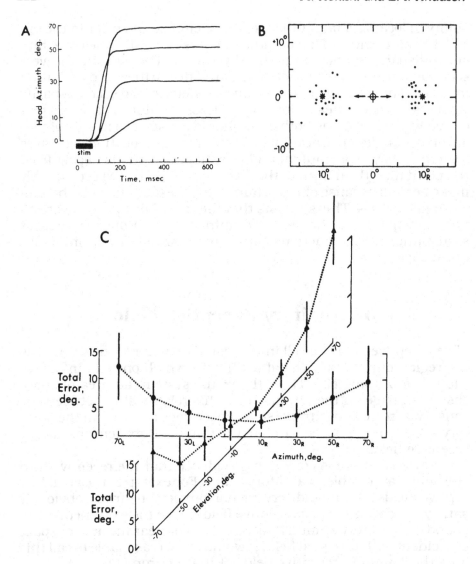

FIG. 7.1. A; time course of head movement. The azimuthal angle of the owl's head is plotted over time, as it orients to sound targets located $10°_R$, $30°_R$, $50°_R$, and 70°R. In each case, the sound stimulus began at time zero and continued for 75 ms. Notice that the sound terminated before the owl moved its head. B; terminal point of head-orienting response around sound targets at 10° and 10°R. The locations of the targets are designated by stars. C; variation in the accuracy of sound localization as a function of target locations. The owl localizes sound most accurately when the target is directly in front of the face. As the target departs from the midpoint of the face, the error of localization increases both in azimuth and in elevation.

determined at that location. To plot the field boundaries, the speaker is gradually moved to locations where the unit barely responds to sound stimuli of either 10 or 30 dB above threshold. The coordinates of these locations constitute the unit's field boundaries (Fig. 7.2) (2). Within a unit's field boundaries there is always an area, called the best area, where the unit shows the lowest threshold. The best area usually corresponds to the azimuthal center of the field, but it need not always correspond to the elevational center.

The anterolateral margin of the owl's midbrain auditory nucleus, MLD, the avian homolog of the inferior colliculus, contains neurons that respond only to an appropriate sound located within a restricted area in space. Their receptive fields are usually vertically elongated ellipses with a slight tilt to the owl's right. In rare cases, a unit's field is a narrow vertical band without any elevational boundary. The size of a receptive field ranges from 7 to 39° in azimuth and 23° to "unrestricted" in elevation. The size as well as the shape of a receptive field is little affected by changes in the intensity of the sound stimulus, although additional weak response areas away from its main field may appear with sufficiently intense stimuli (Fig. 7.3).

As mentioned earlier, we infer from behavioral observations that the auditory system assigns a predetermined set of disparity values to each location in space. The cellular correlate of this scheme would be receptive fields having locations determined by unique sets of binaural disparities. Neurons in the anterolateral area of MLD are binaural and the location of an auditory receptive field depends on binaural disparities; its elevational and azimuthal boundaries are respectively determined by the tuning of the neuron to a specific range of binaural intensity and time differences. Each neuron responds only when both intensity and time disparities simultaneously fall within the range to which the neuron is tuned (8).

A neural representation of space requires that not only individual locations in space but also their relationships must be encoded. There must be some principle to relate one receptive field to another in a manner meaningful for the encoding of space. One manifestation of this pinciple involves excitatory and inhibitory relationships among cells having adjoining receptive fields. This can be demonstrated in the following manner: one speaker is placed within a field and a second one is moved in and out of the field boundary. When the two speakers are within the field, their excitatory effects add up. When the moving speaker is outside the field, sound from it can reduce or eliminate the excitatory effects of the stationary speaker.

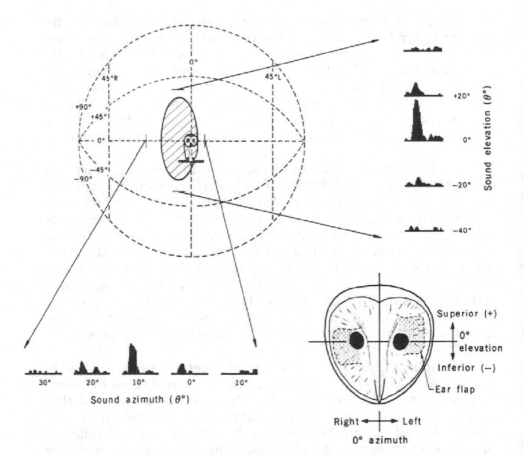

FIG. 7.2. The receptive field of an auditory neuron depicted from the observer's point of view. The owl is shown facing out from the center of the stimulus sphere (dashed globe) and the unit's receptive field (25° in azimuth by 62° in elevation) is projected onto the sphere (diagonally lined area). The unit was located in the owl's left hemisphere; its field dimensions were independent of stimulus intensity. Below and to the right are shown peristimulus-time histograms of the unit's responses to a sound stimulus presented at different locations within its receptive field. The stimulus was a 200 ms noise burst, 20 dB above threshold, delivered once per second. Each histogram is a 500-ms sample and represents 16 stimulus repetitions. Notice the increasing response vigor as the sound source approaches the center of the unit's receptive field. The owl head in the lower right corner illustrates the alignment of the owl in the stimulus sphere and defines the nomenclature used for describing auditory space.

The distribution of excitatory and inhibitory areas is not random, but organized into a center-surround pattern. Thus, the owl's brain treats the hypothetical surface of auditory space in a strikingly similar way to the real sensory surfaces present in the visual and somatosensory systems (5).

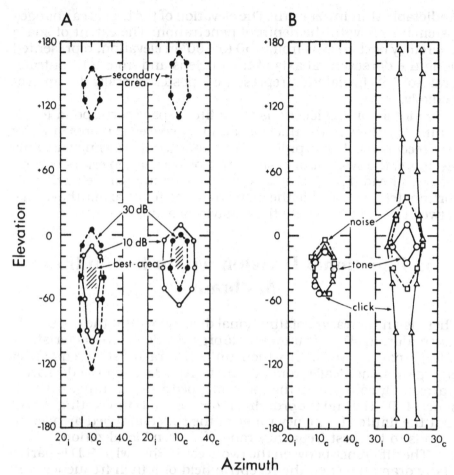

Fig. 7.3. Effect of sound intensity and sound type on the receptive-field plots of limited-field units. A; receptive fields of two units plotted with wide-band noise bursts at 10 dB (open circles) and 30 dB (closed circles) above threshold. The field of the unit on the left expands and the one on the right contracts at the higher sound intensity. Both units exhibited secondary areas behind the owl's head to 30-dB noise bursts. B; receptive fields of two units plotted with noise bursts, CF-tone bursts, and clicks, each presented at 30 dB above threshold. The unit on the left is typical of many units. The unit on the right represents the most extreme case of receptive-field variation owing to sound type.

4. A Neural Map of Auditory Space

The receptive fields of neurons in MLD are represented in a precisely organized fashion across the nucleus (3). In a dorsoventral electrode penetration, all neurons have their best areas within a narrow azimuthal band. Penetration of an adjoining point causes a predictable shift in azimuth. The elevation of the best area changes systematically with the depth of penetration. The extent of space covered by best areas is from +50 to −90° in elevation, represented along the dorsoventral axis of the nucleus and from 60° contralateral to 15° ipsilateral, represented posterolaterally to anteromedially.

Since adjoining locations in auditory space correspond to adjoining locations in the nucleus, auditory space, as represented by unit receptive fields, is projected or mapped onto the nucleus. Furthermore, the systematic distribution of localization errors around the owl's head (Fig. 7.1) seems to have a neural parallel; a disproportionately large area of the map is devoted to the azimuthal range between 0 and 20°, where the behavioral acuity is greatest.

5. Functional Division within the Auditory Midbrain

There is an indication of functional division within the owl's auditory midbrain (4). A clear tonotopic organization is characteristic of a large part of MLD, in which units' characteristic frequencies change systematically from 500 Hz to 10 KHz along the dorsoventral axis. By contrast, in the space-mapped area, the range of CFs is short (4–9 KHz) and the ordering is obscure. It is interesting to note that the limited frequency range of the *space-mapped* area corresponds to the best frequency range for sound localization.

The difference between the two areas in the owl's MLD is particularly dramatic when the receptive field of a high frequency unit from one area is compared with that of the other area. Since high frequencies create greater binaural intensity differences, high frequency units might be expected to exhibit greater spatial selectivity than low frequency units. Many high frequency units in the *tonotopic area* of MLD, however, are insensitive to changes in the location of a sound source; they can be driven equally effectively from any direction. Those tonotopic neurons that exhibit spatial preference tend to have either multiple receptive fields or fields that change with sound intensity. These properties contrast sharply with those of units in the space-mapped area.

6. Concluding Remarks

The map of auditory space shows that the ordering of neuronal arrays according to systematic variation in unit response properties is not confined to those based on the topographical projection of

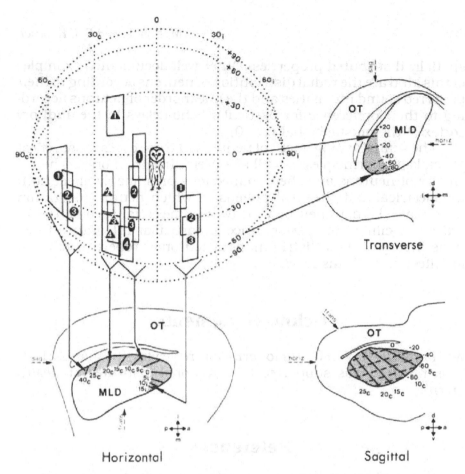

Horizontal   Sagittal

FIG. 7.4. Representation of auditory space in the midbrain nucleus (MLD), as defined by centers of unit best areas. In the upper left, coordinates of auditory space are depicted as a dotted globe surrounding the owl. Projected onto the globe are the best areas (solid-lined rectangles) of 14 units that were recorded in four separate penetrations. The large numbers backed by similar symbols represent units from the same penetration; the numbers themselves signify the order in which the units were encountered and are placed at the centers of their best areas. Penetrations were made with the electrode oriented parallel to the transverse plane, at positions indicated in the horizontal section by the solid arrows. Below and to the right of the globe are illustrated three histological sections through MLD in horizontal, transverse and sagittal planes. The stippled portion of MLD corresponds to the space-mapped region; the remaining portion is the tonotopic region. Isoazimuth contours, based on best-area centers, are shown as solid lines in the horizontal and sagittal sections; isoelevation contours are represented by dashed lines in the transverse and sagittal sections. On each section, dashed arrows indicate planes of the other two sections. Solid, crossed arrows to the lower right of each section define the orientation of the section: a, anterior; d, dorsal; l, lateral; m, medial; p, posterior; v, ventral. The optic tectum (OT) is labeled on each section.

227

spatially distributed properties. Other well-documented examples of this kind are the radial distribution of neurons according to their preferred sound intensities and the linear order of neurons according to their preference for particular echo-delays in the auditory cortex of the mustache bat (9, 10).

The spatial arrangement of neurons in these cases must reflect physiological principles as well as purely structural ones. The geometry of neurons and their connections presumably affect both the electrical and chemical milieus in which neural integration takes place. Perhaps neuronal arrays assume certain geometries in order to facilitate the performance of particular integrative functions. The systematic distributions of neuronal properties may be a manifestation of this need.

Acknowledgments

We thank Dr. D. Van Essen for critically reviewing the manuscript. This research was supported by a National Institutes of Health grant.

References

1. HUBEL, D. H., AND WIESEL, T. N. Receptive fields, binocular interaction and functional architecture in the cat's visual cortex. *J. Physiol., London*, 160: 106–154, 1962.
2. KNUDSEN, E. I., KONISHI, M., AND PETTIGREW, J. D. Receptive fields of auditory neurons in the owl. *Science*, 198: 1278–1280, 1977.
3. KNUDSEN, E. I., AND KONISHI, M. A neural map of auditory space in the owl. *Science*, 200: 795–797, 1978.
4. KNUDSEN, E. I., AND KONISHI, M. Space and frequency are represented separately in auditory midbrain of the owl. *J. Neurophysiol.*, 41: 870–884, 1978.
5. KNUDSEN, E. I., AND KONISHI, M. Center-surround organization of auditory receptive fields in the owl. *Science*, 202: 778–780, 1978.
6. KNUDSEN, E. I., BLASDEL, G. G., AND KONISHI, M. Sound localization by the barn owl (*Tyto alba*) measured with the search coil technique. *J. Comp. Physiol.*, 133: 1–11, 1979.
7. KNUDSEN, E. I., AND KONISHI, M. Mechanisms of sound localization in the barn owl (*Tyto alba*). *J. Comp. Physiol.*, 133: 13–21, 1979.
8. MOISEFF, A. AND KONISHI, M. Neuronal and behavioral sensitivity to binaural time differences in the owl. *J. Neurosci.*, 1: 40–48, 1981.

9. SUGA, N. Amplitude-spectrum representation in the Doppler-shifted-CF processing area of the auditory cortex of the mustache bat. *Science*, 196: 64–67, 1977.

10. SUGA, N., AND O'NEILL, W. E. Neural axis representing target range in the auditory cortex of the mustache bat. *Science*, 206: 351–353, 1979.

11. WICKELGREN, B. G. Superior colliculus: some receptive field properties of bimodally responsive cells. *Science*, 177: 69–71, 1971.

Chapter 8

Cortical Auditory Area of *Macaca mulatta* and Its Relation to the Second Somatic Sensory Area (Sm II)

Determination by Electrical Excitation of Auditory Nerve Fibers in the Spiral Osseous Lamina and by Click Stimulation

C. N. Woolsey and E. M. Walzl

Department of Neurophysiology and Waisman Center, University of Wisconsin, Madison

1. Introduction

The study here reported was carried out more than 35 years ago, when the method of electrical stimulation of nerve fibers in the exposed spiral osseous lamina, first applied to the cat by Woolsey and Walzl (43, 44) was employed to define topological relations between the cochlea and the cortical auditory field in the monkey [Walzl and Woolsey (33); Woolsey and Walzl (45)]. Only a general diagram of the results has ever been published [Woolsey and Fairman (42); Walzl (32); Woolsey (37); Davis (11); Woolsey (41)]. The present account fully reports the data collected and undertakes to relate the findings to more detailed results now available on the cat [Woolsey (41); Reale and Imig (29)] and to other auditory studies on primates [Ades and Felder (1); McCulloch et al. (23); Licklider and Kryter (22); Pribram et al. (28); Kennedy (20); Woolsey (41); Neff (25); Bailey et al. (4); Brugge and Merzenich (7); Merzenich and Brugge (24); Imig et al. (19)].

In other studies, the auditory area of the monkey was mapped by click stimulation [Ades and Felder (1); Pribram et al. (28)] or with tonal stimuli [McCulloch et al. (23); Licklider and Kryter (22); Kennedy (20); Neff (25)]. Comparative aspects of cortical organization have been aided by studies on the second somatic sensory area [Woolsey (35, 36)] and by a study of this area in the squirrel monkey [Benjamin and Welker (5)].

2. Material and Methods

Fourteen *Macaca mulatta* monkeys of both sexes, ranging in weight from 1.3 to 8.2 kg, were used.

All animals were deeply anesthetized with pentobarbital sodium (30 mg/kg initially, with additional amounts as required), except for one animal (M-12), which received "Dial." While data were being collected, the anesthesia was kept at a level sufficient to reduce spontaneous electrical activity to a minimum and keep the evoked responses as constant as possible.

In all experiments, in which cochlear nerve fibers were stimulated, the right cochlea was dissected by Walzl after opening the tympanic bulla by a mastoid approach.

For exposure of the cortical auditory fields on the ventral and dorsal walls of the Sylvian fissure, two brain dissections were necessary. These are illustrated in Fig. 8.1a. Before removing cortex above or below the Sylvian fissure, the large arteries and veins traversing the area to be ablated were ligated with fine silk close to the

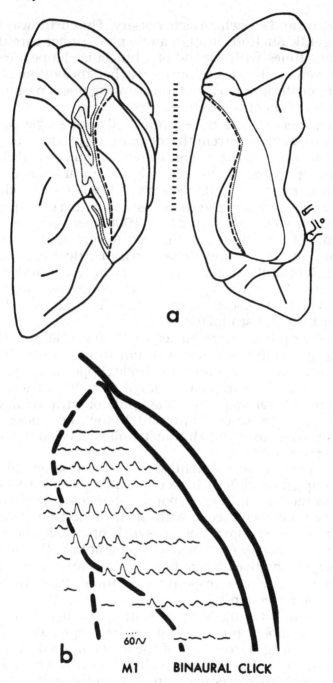

FIG. 8.1. (a) Resections of frontoparietal opercula to expose the lower bank of the Sylvian fissure (left) and of the temporal lobe to expose the upper bank of the Sylvian fissure (right); millimeter scale between brain drawings; (b) binaural click map of the lower bank of the Sylvian fissure of monkey M1. Surface positivity up.

Sylvian fissure and elsewhere as necessary. The cortex was then re-moved by gentle suction through a fine pipet under careful visual control, sometimes with the aid of a binocular loupe. Great care had to be taken, as the dissection approached the bottom of the Syl-vian fissure posteriorly, to avoid interrupting fibers passing to the upper or lower wall of the fissure.

Cochlear nerve fibers were stimulated with single condenser discharges of weakest strength adequate to produce a good re-sponse at the center of the response area. Stimuli were delivered to cochlear nerve fibers at the edge of the spiral osseous lamina through fine, enameled, stainless steel, bipolar electrodes sepa-rated by 0.5–0.7 mm and placed in position with a special manipu-lator. It is estimated that around 500–1000 fibers near the cathode were activated by the electrical stimulus. Reversing the direction of current through the electrodes changed the cortical locus of best re-sponse. Click stimuli, capable of exciting the whole auditory area, were produced by activating with single condenser discharges a 10-in. loudspeaker placed about 1 foot from the head (in the midline for binaural stimulation).

Cortical responses were picked up with an "active" electrode consisting of a saline-moistened cotton thread protruding from steel tubing and an "indifferent" electrode clipped to the skin of the brow. The "active" electrode was moved over the cortical surface, usually in millimeter steps, by a manipulator with millimeter rec-tangular coordinate scales. To prevent drying, the exposed surface of the cortex was covered with mineral oil. This also tended to re-strict the recording site.

The responses were amplified by a three-stage, resistance-capacity coupled amplifier with a time constant of 0.1 s (Johnson Foundation model No. 39). Responses were visualized on a 3-in. RCA cathode-ray oscilloscope with linear deflection, modified to permit single sweep operation in synchrony with the stimulus [Talbot (31)]. Records were made by photographing the cathode-ray trace on supersensitive 35 mm film with a Leica camera. Detailed records were also kept of response amplitudes and wave forms by sketching them freehand.

Illustrations were made by mounting photographs of the re-sponses, or symbols proportional to the amplitudes of the re-sponses, on enlarged drawings of the areas studied, in positions corresponding to the points at which the records were taken. The drawing in each case was made from a photograph of the actual brain studied. Recording sites were charted on graph paper and all fissures and other anatomical landmarks were plotted, so that the chart and brain photograph could be related.

3. Results

3.1. Cortical Area Responsive to Clicks

The area of the *lower bank* of the Sylvian fissure responsive to strong click stimuli was mapped in four animals (M1, M2, M4 and M7) and partially mapped in another (M3), while cortex of the upper bank was examined with click stimuli in two cases (M8 and M9).

In M1 (Fig. 8.1b), the area responsive to *binaural stimulation* occupied the lower bank of the Sylvian fissure from about two millimeters ahead of its posterior end to four millimeters rostral to the caudal end of the insula. The responses diminished in amplitude laterally and only posteriorly did they reach the lip of the Sylvian fissure. Responses were also recorded from the cut fibers to the upper bank of the Sylvian fissure. All responses were initially surface positive, with onset latencies of approximately 12–14 ms, peak latencies of 22–28 ms, durations from 25 to 35 ms and maximal amplitudes of 250 mV. In M8 (Fig. 8.5b) and M9 (Fig. 8.5c), the *upper bank* of the Sylvian fissure was explored, while both ears were stimulated with clicks. In M8, the area of response extended 4–5 mm outward from the depth of the fissure, while in M9 the area was narrower and the responses were smaller.

In M4 (Fig. 8.4b, lower left), the response area to click stimulation of the *contralateral* ear was similar in extent to the binaural response area of M1, reaching the lip of the Sylvian fissure only caudally.

In M2 (Fig. 8.2b, right), the area responsive to click stimulation of the ipsilateral ear was more extensive, reaching the caudal end of the lower bank of the Sylvian fissure and extending everywhere to the lip of the sulcus and even onto the superior temporal gyrus near the caudal end of the Sylvian fissure. Unfortunately, the superior temporal gyrus was not explored more rostrally. Good responses were also recorded from the cut fibers to the upper bank of the sulcus and from the caudal end of the insula. In two other animals, M7 and M3, ipsilateral click responses were recorded from the lower bank. The area in M7 (Fig. 8.3) was somewhat smaller than in the animals already considered. In general, all responses in this animal were smaller in size and the animal may have been too deeply anesthetized. The response area to ipsilateral click in M3 (Fig. 8.4a) was only partially mapped but, in the portion defined, extended well out toward the lip of the Sylvian sulcus. Death of the animal prevented completion of the map, but since the fibers to the caudal part of the area had been partially undercut in removing the upper bank, it

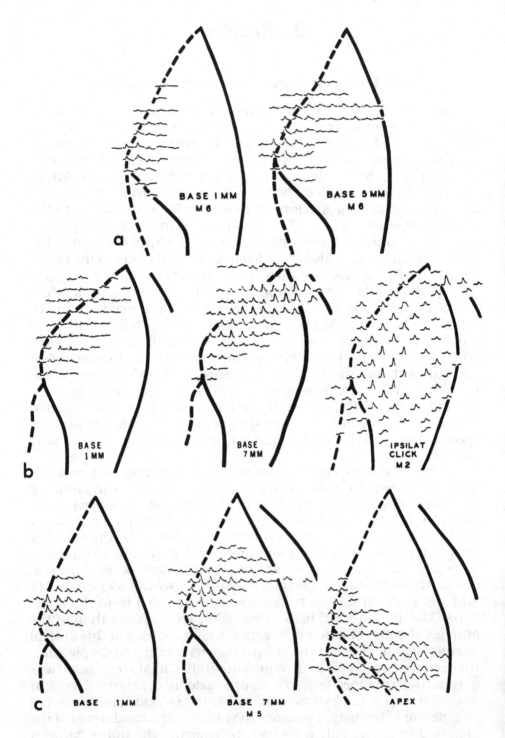

FIG. 8.2. (caption on facing page)

probably would not have been possible to record responses from the caudal part of the area.

3.2. Topical Projection of Nerve Fibers from Local Regions of the Cochlea to the Cerebral Cortex

In eight experiments (M2, M3, M4, M5, M6, M7, M12 and M14), cochlear nerve fibers in the exposed spiral osseous lamina were stimulated with single electric shocks.

In M6, the stimulating electrodes were applied at sites 1 and 5 mm from the basal end of the spiral lamina. The distribution of responses on the lower bank is shown in Fig. 8.2a. For each site stimulated, the responses were largest in the depth of the fissure caudal to the posterior end of the insula. Responses to stimulation of the 5 mm point extended more posteriorly and laterally as far as the lip of the Sylvian fissure.

In M2 (Fig. 8.2b), sites 1 and 7 mm from the basal end of the cochlea were stimulated. The response area for the 1 mm site extended all along the bottom of the sulcus from the insula to the caudal end of the Sylvian fissure. The response area for the 7 mm point was more extensive and the responses were of greater amplitude. There may be two foci in this response area, one more medially and the other posterolaterally at the lip of the Sylvian fissure. A more obvious second focus of high amplitude at the lip of the Sylvian fissure is evident in the ipsilateral click map. Surface positive responses were also recorded from the cut fibers to the upper bank for both cochlear sites stimulated. A slower cathode-ray sweep was used during stimulation of the 1 mm basal point than for the 7 mm basal point and for click stimulation.

In M5 (Fig. 8.2c), three cochlear sites were stimulated: 1 mm, 7 mm basal points and apex. The distributions of evoked potentials on stimulation of the 1 and 7 mm sites differed from the results shown in Fig. 8.2b for the same cochlear sites in being less extensive and not extending as far caudalward. Again, however, there appears to be a second focus for the 7 mm site on the superior temporal gyrus somewhat more rostrally located than in Fig. 8.2b. The response area for stimulation of the apex of the cochlea was located largely lateral to the caudal 5 or 6 mm of the insula and reached

FIG. 8.2. Responses to focal electrical stimulation of cochlear nerve fibers in the spiral osseous lamina: (a) M6 and (1 mm and 5 mm basal points): (b) M2 (1 and 7 mm basal points and ipsilateral click): (c) M5 (1, 7 mm basal points and apex).

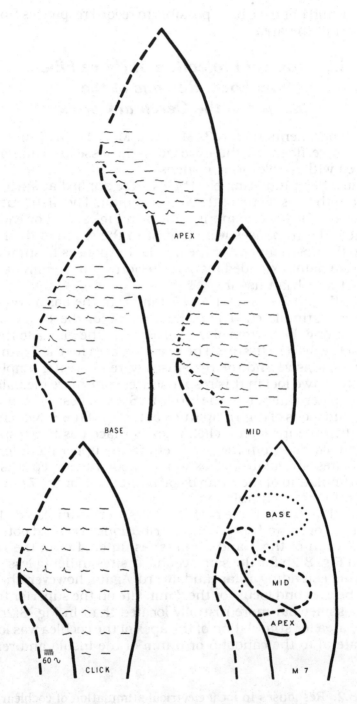

FIG. 8.3. Responses to stimulation of basal middle and apical turns of the spiral osseous lamina in M7 and the effects of click stimulation. Summary diagram at lower right. Time signal at lower left, 60 cycles per second.

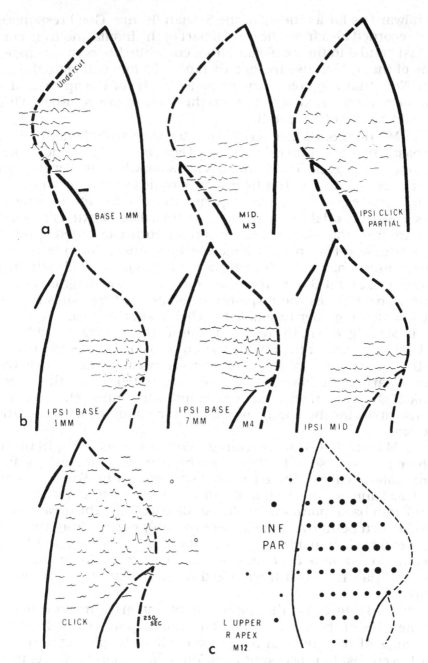

FIG. 8.4. (a) Fibers to posterior part of lower bank of Sylvian fissure have been undercut in M3. Incomplete response area for 1 mm basal point and for ipsilateral click map. Map for middle turn site satisfactory; (b) ipsilateral response areas for 1 mm, 7 mm basal sites and for middle turn site and click map in M4; (c) map of upper bank of Sylvian fissure and lower inferior parietal cortex to stimulation of right apex in M12. Dots are proportional to the amplitudes of surface positive responses.

lateralward as far as the lip of the Sylvian fissure. Good responses were recorded also from the caudal part of the insula and from cortex just caudal to the insula and from cut white fibers to the upper bank of the Sylvian fissure for a distance of 5 mm caudal to the insula. Note that the axes of the response areas for the apical and 7 mm sites are approximately at a right angle to one another. This also can be seen in Fig. 8.3.

In M7 (Fig. 8.3), three cochlear sites were stimulated, one on the basal turn, 2–3 mm from the basal end, one on the middle turn and one at the apex. The responses are somewhat smaller than in other experiments, with little evidence of negativity following the initial surface positivity. The response areas for the three cochlear sites overlap caudal to the insula, but their lateral extensions fan outward from the overlap area. The area of cortex activated by these three sites and the area of click evoked responses does not fill the whole caudal portion of the lower bank; the most caudally situated response is about 6 mm from the caudal end. The diagram at the lower right outlines the response areas for the three sites stimulated, which together resemble the click response area.

In M3 (Fig. 8.4a), the results of stimulating the right cochlea at its basal end and in the middle coil are shown. The response area for the 1 mm point probably is incomplete, because the fibers to the caudal part of the lower bank were interrupted, when the upper bank of the Sylvian was removed as indicated (undercut). The response areas for the 1 mm and middle turn points show very little overlap.

In M4 (Fig. 8.4b), all recordings were made from the right hemisphere in response to ipsilateral cochlear nerve stimulations and contralateral clicks. Three cochlear sites were stimulated, two on the basal and one on the middle turn. The response areas for the 1 and 7 mm basal points were almost identical, but the area for the middle turn point overlapped these two only partly. From the contralateral click map (lower left), one would have expected the 1 mm map to extend more caudad, since the fibers to the lower bank apparently had not been interrupted during ablation of the upper bank.

In M14 (Fig. 8.5a), the amplitudes of responses are represented by the sizes of the solid circles. (This dot map was made from the drawings of responses in the protcol, when the photographic records were lost in processing.) The 7 mm basal point projects to an area very like the 7 mm area in Fig. 8.4b. The area of response to stimulating the right apex covers in the ipsilateral (R) hemisphere the cortex of the lower bank from the insula to the lip of the Sylvian fissure and extends onto the superior temporal gyrus. The area also includes the caudal cortex of the lateral wall of the inferior limiting sulcus of the insula and a few points caudal to the insula at the bottom of the Sylvian fissure. The right apex also projects to the upper bank of the left Sylvian fissure as illustrated at the right, where the left upper bank has been drawn as if it were the right upper bank, in order to relate the data from lower and upper banks as a single response area.

In M12 (Fig. 8.4c), dots also represent responses recorded from the upper bank of the left Sylvian fissure on electrical stimulation of the left cochlear apex. This was the most extensive projection of auditory input to the upper bank of the Sylvian fissure seen in these experiments. The area occupied the whole upper bank from 4 mm rostral to the caudal end of the insula almost to the caudal end of the upper bank. Responses were also recorded from the inferior parietal lobule on the lateral surface of the hemisphere, as indicated. In another animal (M13), not illustrated, small responses to click stimulation were recorded from the inferior parietal lobule as far caudad as the upper end of the Sylvian fissure and as far rostrad as the lower end of the intraparietal sulcus and from cortex between the Sylvian and the intraparietal sulci.

3.3. The Second Somatic Sensory Area

In six animals (M5, M7, M8, M9, M10 and M13), observations were made on the second somatic sensory area (S II). Since these studies were carried out after the auditory data had been collected, the time devoted to them was necessarily limited and detailed somatotopic relationships to displacement of hairs were not defined.

Figure 8.5b illustrates the responses recorded from the left upper bank of the Sylvian fissure of M8, when fingers and toes were stimulated strongly by the lever arm of the tactile stimulator. Responses to stimulation of the right fingers were recorded from the depth of the sulcus to the lip of the Sylvian fissure and from near the caudal end of the superior bank to about 4 mm anterior to the caudal end of the insula. Responses to stimulation of the toes were limited to the deeper part of the upper bank to an area also activated by binaural click stimuli. Although bar stimulation was used to map the projection areas for fingers and toes, it also possible in this animal to evoke responses under lighter anesthesia to tactile stimulation of hairs on the surface of the hand.

Figure 8.5c (M9) shows similar results recorded from the upper bank of the left Sylvian fissure to bar stimulation of the proximal phalanges of fingers II-V and toes II-V. The response area to finger stimulation covers the entire upper bank from about 2 mm ahead of the caudal end of the insula to the posterior end of the Sylvian fissure. Again the toe area was restricted to the deeper portion of the

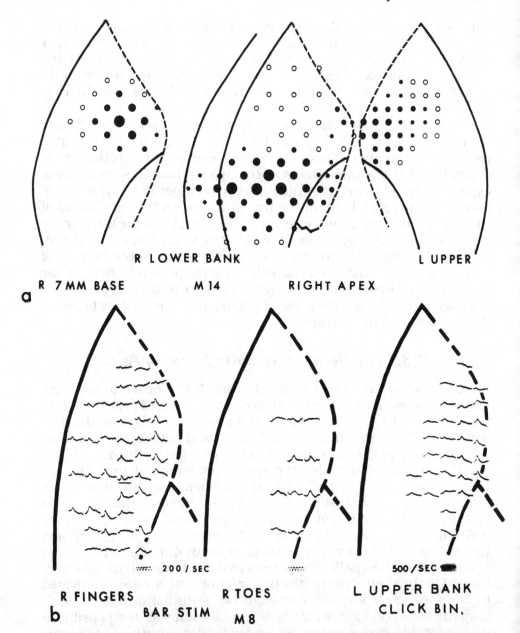

upper bank and was also overlapped by the area responsive to binaural click stimuli. Movement of hairs on the lateral aspect of the little finger evoked a response about half the amplitude of the bar elicited responses. Under light anesthesia, responses were recorded to ipsilateral stimulation of both the arm and the leg.

For a short period after a supplementary dose of Nembutal, re-

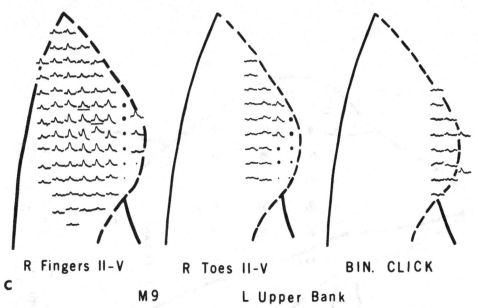

R Fingers II–V   R Toes II–V   BIN. CLICK

C

M9   L Upper Bank

Fig. 8.5. (a) Response areas on the right lower bank of the Sylvian fissure to cochlear nerve stimulation of a 7 mm basal point and of the apex in the right cochlea of M14. At the right the response area of the left upper bank of the Sylvian fissure to stimulation of the right apex is illustrated as if in the right hemisphere. Dots are proportional to the amplitudes of surface positive responses. The lateral wall of the right inferior limiting sulcus of the insula was exposed by removal of insula. (b) Responses to stimulation of fingers and toes with bar of tactile device as recorded from left upper bank of Sylvian. Binaural click responses at right. (c) Similar results from left upper bank of M9 to pressure applied to skin and to binaural clicks.

sponses were suppressed in both S I and S II. The responses returned first in S I and then in S II. In the cat, where the latency of the S II response is shorter than that of the S I response, the opposite is true. Thus, both in monkey and cat, the longer latency response is more easily depressed by the anesthetic.

Figure 8.6, at the right, shows by the sizes of the dots the relative amplitudes of responses recorded in the second somatic area (S II) on the upper bank of the left Sylvian fissure of MI3 on tapping the right thumb. The response area does not extend as far caudad as the finger area in Fig. 8.5c. Responses to stimulation of the right foot were obtained in the area labeled "leg" at the bottom of the Sylvian fissure.

The evoked potentials plotted at the left in Fig. 8.6 were obtained on brush stimulation of the right upper lip. Two groups of responses can be seen, those in S I and the posterior parietal gyrus

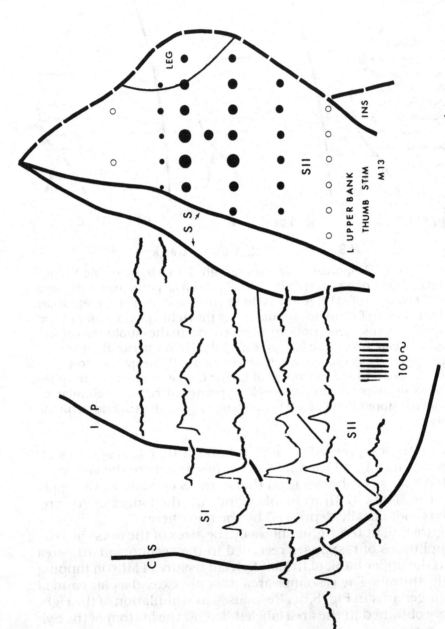

FIG. 8.6. Right, distribution of responses in S II (proportional to sizes of the dots) to bar stimulation of right thumb in M13. Leg area deep within the Sylvian fissure. Left, responses evoked in S I and S II by brush stimulation of the right upper lip.

above the narrow line and those below this line, in what we believe to be part of the face area of S II. The latencies are shorter in S I (5.6–8.0 ms) than in S II (around 12 ms).

In Fig. 8.7a are illustrated responses recorded from the little finger area of the postcentral gyrus (S I) to tapping the little finger of M9. The latency to onset of the sharp rise of the response is approximately 8.3 ms, peak latency is about 13.6 ms and the duration of the positive deflection is around 8.3 ms, while the negative deflection lasts about 15.5 ms.

Figure 8.7b illustrates two responses recorded from the second somatic sensory area (S II) at the site of the lower of the two underlined responses in Fig. 8.5c, to the same stimulation of the little finger. The latency to the onset of the fast rise of the surface positive deflection is 12 ms and to the peak of the response is 19 ms; the duration of the positive deflection is about 20 ms, while the negative

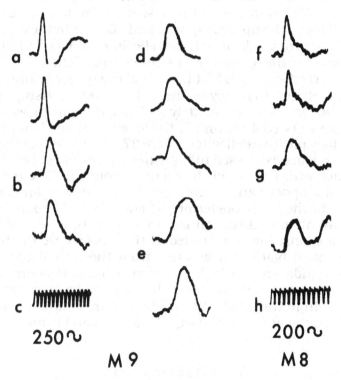

250∿   200∿

M 9   M 8

FIG. 8.7. (a) Responses recorded in S I on stimulation of the contralateral little finger; (b) responses in S II to the same stimulus; (d) responses recorded from lower underlined response site in Fig. 8.5c, when pressure was applied to the contralateral hand; (e) responses in S II to tapping contralateral foot; (f) responses in S I to tapping palmar aspect of contralateral digits II–V; (g) responses in S II to same stimulus used in f; (c) time signal for M9 responses; (h) time signal for M8 responses.

deflection lasts approximately 24 ms. Figure 8.7d illustrates three responses recorded in S II at the other site underlined in Fig. 8.5c to pressure applied to the contralateral hand. The latencies of onset of the rapid rise of the surface positive deflections and of the peaks of the responses are the same as for the two responses of Fig. 8.7b. The responses are mainly surface positive and their durations are slightly longer than those of Fig. 8.7b—about 24 ms. The responses of Fig. 8.7e were induced by tapping the contralateral foot. The latencies to onset are around 15.5 ms and the peak latencies are 32 and 34 ms. The durations of the positive deflections are 36 and 32 ms.

In Fig. 8.7f (M8), two responses recorded from the postcentral gyrus near the lateral end of the intraparietal sulcus (S I) to tapping the palmar surfaces of the proximal phalanges of digits II-V of the right hand are shown. The latency to onset of the sharp rise in the surface positive deflection in both responses is 11 ms and to the peak 15.5 ms. The duration of the positive deflection was approximately 24.5 ms in the upper response and 20.0 ms in the lower one. The negativity is of small amplitude in both responses. In Fig. 8.7g, two responses to the same site of stimulation of the contralateral hand were recorded from S II at the site of the response underlined in Fig. 8.5b. Here the latency to onset of the sharply rising surface positive deflection is approximately 13.5 ms, the time to peak of the upper response is 15.4 ms and of the lower 19.6 ms. Durations of the essentially positive deflections were 27 and 22 ms, respectively.

In Fig. 8.8 are expressed in figurines the results of tapping the body surface with the bar of the tactile stimulator. The recording area is on the upper bank of the right Sylvian fissure and extends outward onto the lower posterior parietal lobule. Responses were obtained at all points examined to stimulation of the hand; strongest responses were recorded at the sites where the hand is marked in solid black and weakest from the caudal two points where the hands are marked by hatched lines. Responses were evoked by stimulation of the leg at the four caudal sites (up in the figure), best from the middle two of these four. This area agrees in position with the areas for the leg in Fig. 8.5b and c, and Fig. 8.6.

4. Discussion

4.1. The Auditory Cortex as Defined by These Experiments

When these experiments of stimulating electrically cochlear nerve fibers to study localization in the auditory region of the cerebral cortex of the monkey were carried out, the monkey's auditory area

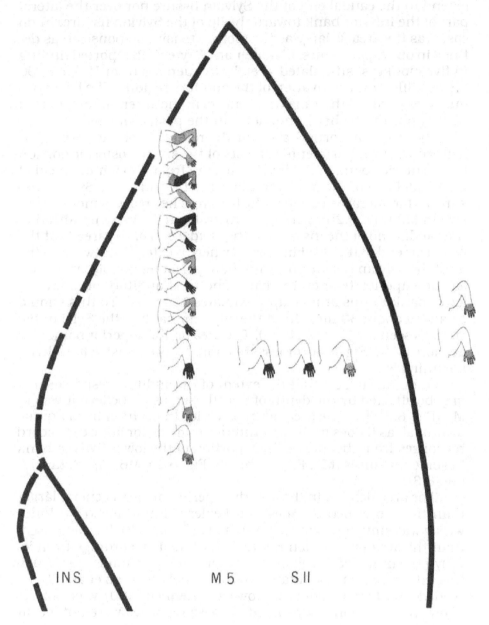

FIG. 8.8. Figurine map for portions of arm and leg areas of S II on right superior bank of Sylvian fissure of M5. Caudal at top.

had recently been studied with click stimuli by Ades and Felder (1) and with tonal stimuli by Licklider and Kryter (22). The area defined by click was situated on the inferior bank of the Sylvian fissure lateral and caudal to the posterior end of the insula. It did not extend to the caudal end of the Sylvian fissure nor over the lateral part of the inferior bank toward the lip of the Sylvian fissure. In no case was the area as large as the more extensive response areas defined in our experiments. Licklider and Kryter (22) reported finding in five monkeys, stimulated at eight frequencies from 125 to 8000 cycles, differential activation of the auditory region: "The lower frequencies produce their maximal effects in the antero-lateral part of the region, the higher frequencies in the postero-medial."

The maximal cortical area for the reception of auditory input defined in our experiments consists of the entire posterior portion of the inferior bank of the Sylvian fissure from 4 to 5 mm ahead of the caudal end of the insula to the posterior end of the Sylvian fissure, as far lateral as the lip of the fissure. The area was most extensive in M2 (Fig. 8.2b), where the rostral margin is 5 mm ahead of the caudal end of the insula and the caudal end of the area is at the posterior end of the Sylvian fissure where it extends outward to the lip of the Sylvian fissure and onto the superior temporal gyrus as far as the superior temporal sulcus. The total auditory area on the lower bank in this animal approximates 130 mm^2. To this should be added about 30 mm^2 from the superior bank of the Sylvian fissure, as seen in M14 (Fig. 8.5a). The area on the superior bank may be more (M12, Fig. 8.4c) or less (M9, Fig. 8.5c) extensive in individual animals.

Our data suggest that the extent of the auditory response area may be affected by the depth of anesthesia, as we believe it was in M7 (Fig. 8.3). But the area perhaps varies in position from animal to animal, as it does in the cat, thus accounting for failure to record responses from the most caudal portion of the lower Sylvian bank in some instances (M1, Fig. 8.1b; M5, Fig. 8.2c; M6, Fig. 8.2a; M7, Fig. 8.3).

The area defined in these early experiments is in general larger than that described by Ades and Felder (1) in *Macaca mulatta*, which was similar in size to the area in M7 (Fig. 8.3). It is also larger than the area from which Kennedy (20) made recordings from locally strychninized areas of cortex with tonal stimuli. It is much less extensive than the total area from which Pribram et al. (28) recorded auditory responses. However, Kennedy (20) was able to show that the more widespread of these reponses were not locally generated.

4.2. More Recent Studies
of Auditory Cortex in Primates

More recent studies of the auditory region in primates have extended the cortical area from which recordings could be made beyond the limits defined by our early experiments. This appears to be the case in *Saimiri* (17) and *Aotus* (14). Merzenich and Brugge (24) did not explore with single unit recordings the whole superior temporal plane posterior to the caudal end of the insula, but they did report finding a rostral area (R) anterior to the low frequency focus of A I. Imig et al. (19), in their microelectrode study of *Aotus*, reported finding two areas rostral to A I (R and AL) and an area (PL) caudal to A I. In addition, they labeled two areas (CM and RM) situated medially, in cortex adjacent to the insula, of which the caudal one may be homologous with A II of the cat. Brugge (Chapter 3, this volume) describes A I, PL and R for *Galago crassicaudatus*.

Area PL of *Aotus* appears to be homologous with area A, the anterior area of the cat, so-named by Knight (21), but originally called the suprasylvian fringe area (SSF) by Woolsey (39) after Rose's (30) anatomical description. This area is situated caudally in *Aotus*, because the orientation of the auditory area in primate brains has been greatly altered by the development of the temporal lobe during evolution, as the temporal lobe swings forward beneath and lateral to the insula. [For chimpanzee auditory area see Bailey et al. (4) and Fig. 5.4 in Woolsey (41)].

There is perhaps evidence for area PL in our monkey M2. In this animal, both the click map and the responses to stimulation of the 7 mm basal point show two maxima. The second is located caudally near the lip of the Sylvian fissure in each map. These second foci of maximal response could represent in *M. mulatta* the high frequency end of the PL area. The low frequency part of this area would then lie on the superior bank of the Sylvian fissure, as in Figs. 8.5a and 8.4c.

In some of our animals, the caudal end of the lower bank of the Sylvian fissure did not respond to cochlear nerve stimulation, as noted above. This suggests that activation of PL in *M. mulatta* may be adversely affected at times by the depth of anesthesia. No evidence exists for a depressing effect of anesthesia on the anterior area A of the cat, but it is well-known that anesthesia may affect comparable areas in different species quite differently. For example, S II of the cat may be less depressed initially by Nembutal than S I, while the reverse is true for *M. mulatta*.

There is no evidence from our cochlear nerve stimulation experiments, nor from our click stimulations, for auditory response areas in the temporal lobe of the monkey rostral to A I. These areas apparently are homologous with the caudal areas, V, P and VP, of the cat described by Reale and Imig (29). In the cat, we know from experience that these areas are more readily depressed by Nembutal anesthesia than are A I and A. It is probable that deeper anesthesia suppressed these areas in our experiments on *M. mulatta*. A less deep anesthesia was used by Merzenich and Brugge (24) and by Imig et al. (19).

4.3. Homologies of Auditory Areas in Cat and Monkey

We shall now consider the possible parallels between the organization of the auditory region of the cat, based on the recent paper of
Reale and Imig (29; see also Chapter 1 of this volume), and the organization of the auditory region of primates (Fig. 8.9). Although
the diagram for the monkey resembles the brain of *M. mulatta*, the
evidence on which it is based is largely derived from study of the
paper by Imig et al. (19) on *Aotus* and from earlier data of Hind et al.
(17) on *Saimiri* and of Earls and Hirsch (14) on *Aotus*, as illustrated
by Woolsey (41).

Figure 8.9 compares the auditory areas of the cat with those for
monkeys. In the cat, six well-defined areas are illustrated in Fig. 8.9
(A, A I, A II, V, P and VP) and there appears to be seventh, VV. In
the monkey also, seven areas are shown listed in the same sequence
as those for the cat (Fig. 8.9). These are PL, A I, A II, RM, RL, AL and
AM. The letters a and b in individual areas mark apical and basal
ends of the cochlear representations.

Evidence for the seven areas in the cat is based on the data
presented by Reale and Imig (29). All areas are clearly defined by
them except for the frequency representation in A II, which is based
on earlier evoked potential data from our own studies (13, 44), and
for representation in VV. This area is on the rostral bank of the posterior ectosylvian sulcus, an area not often studied. However, in one
experiment, illustrated by Fig. 4 of Reale and Imig (29) and by Fig.
26 of Imig and Reale (18), high frequencies were represented low
down on the anterior bank of this fissure. This location for high frequency representation fits well with the high frequency representation near the insula in *Aotus*, illustrated by Fig. 8 of Imig et al. (19).
This area we are labeling AM (anteromedial). Three other areas, in
addition to AM, are illustrated as existing rostral to A I on the lower
bank of the Sylvian fissure in monkeys in the region labeled R by
Merzenich and Brugge (24) in *M. mulatta* and R and AL by Imig et
al. (19). We have subdivided R and AL into medial (RM and AM) and
lateral areas (RL, AL) based on their detailed frequency organizations. RM includes area RM of Imig et al. (19). RM is organized with
high frequency representation near the insula and resembles area
V rather than area P of the cat.

Area RL, considered homologous with cat's area P, is not identifiable in the study of Imig et al. (19), where the lateral surface of
the superior temporal gyrus was not mapped as far laterally as we
believe it should have been. However, a suggestion of the high frequency part of this area is seen in their Fig. 7D. Figure 5.6 of Woolsey (41), based on the work of Earls and Hirsch (14), indicates that
the lateral border of A I, especially at the low frequency end, is well

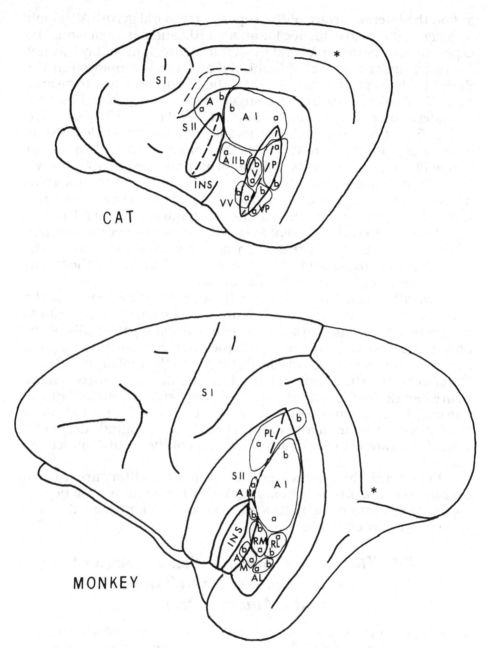

FIG. 8.9. Homologies of auditory areas in cat and monkey, based on current information as reviewed in text: Cat: A (anterior); A I; A II; V (ventral); P (posterior); VP (Ventroposterior); VV (ventroventral). Monkey: PL (posterolateral); A I; A II; RM (rostromedial); RL (rostrolateral); AL (anterolateral); AM (anteromedial). The asterisks mark the approximate locations of the center of gaze at the boundary of V I and V II in visual cortex.

out on the lateral surface of the superior temporal gyrus. Although this figure gives no evidence for an area RL, such an area would be expected to lie between A I and EP of that figure. It probably was not activated under the level of Nembutal anesthesia employed in the Earls-Hirsch experiments. The EP of their map we believe to be area AL of this account. Both these suggestions are supported by a reconsideration of the results on *Saimiri* (Hind et al., 17) illustrated in Fig. 5.5 of Woolsey (41). In this figure, the area labeled SSF is clearly not correctly labeled. SSF is the homolog of area A in the cat and of PL in monkeys and would be situated at the left in Woolsey's Fig. 5.5, perhaps in the area labeled "A II, 0.5" and the adjacent portion of the 21 KHz section of A I. The area labeled SSF clearly has its low frequency end adjacent to the low frequency end of A I and its high frequency end is adjacent to the high frequency end of the area labeled EP. These two areas then are now considered to be the equivalents of areas P and VP of the cat or of RL and AL of the terminology used in this paper for the monkey.

One other point may be made in support of the view that the areas on the superior temporal plane and adjacent superior temporal gyrus are analogous to the areas in the cat on the walls of the posterior ectosylvian sulcus and on the posterior ectosylvian gyrus. Several studies of retrograde degeneration in the thalamus, after lesions involving the temporal lobe of the monkey rostral and lateral to area A I (2, 3, 8, 9) and the walls of the posterior ectosylvian sulcus and the posterior ectosylvian gyrus (12, 26, 30) of the cat, show that these lesions in monkey and cat result in retrograde degeneration of the caudal pole of pars principalis of the medial geniculate body.

Figure 8.9, then, indicates that the main auditory areas of cat and monkey are probably homologous. Further studies will be necessary to confirm or deny this suggestion, since present data are not sufficiently complete.

4.4. The Second Somatic Sensory Area of the Monkey and Its Relation to the Auditory Area

In several animals used for auditory studies, the second somatic sensory area was explored as a terminal project. Time did not permit a very detailed study of the relations of the cutaneous surface to the cortex of the superior bank of the Sylvian fissure. From the observations made, however, it appeared that leg was most deeply represented on the superior bank, with trunk and proximal arm deep in the fissure and with face more rostral and less deep in the fissure. This general layout has been confirmed by Benjamin and

Welker (5), Whitzel et al. (34), by Burton and Robinson (10) and by Friedman (15) in much greater detail and with additions. The orientation of S II to S I and to the insular region in the monkey is similar to that described originally for the monkey (35, 36) and for the dog (27), but not to that reported for the cat by Haight (16).

The results reported here in detail were the first data recorded from the second somatic sensory area of any primate brain (35, 36). The upper bank of the Sylvian fissure was explored to test the hypothesis concerning the formation of the Sylvian fissure discussed by Woolsey and Walzl (44) in their paper on the cat. It proved difficult to evoke activity in S II of the monkey under Nembutal anesthesia and it was not possible to define in detail the topographic relations between body surface and brain by movement of hairs. Stronger stimulation of body surface generally was required, but under certain conditions of lighter anesthesia, hair movement was effective and it was possible to record responses from stimulation of both sides of the body, as had been observed in the cat (35). The observations made on the second somatic sensory area confirmed the view developed by Woolsey and Walzl (44) concerning the homologies of cortical areas in cat and monkey.

Acknowledgments

This work was carried out in the Department of Physiology, The Johns Hopkins University School of Medicine, between October 24, 1942 and November 11, 1944. Dr. Walzl died on August 10, 1950 at the age of 40 (40).

Gratitude is here expressed to the wives of the authors, Harriet Woolsey and Florence Walzl, for the care which they devoted to keeping detailed protocols of the auditory experiments and to Dr. Ging-Hsi Wang for assistance in some of the second somatic sensory area experiments. Thanks is also extended to Evadine Olson for typing and to Terry Stewart and Shirley Hunsaker for photographic assistance. Aided in part by research grants NB-03641, NB-06225 and HD-03552.

References

1. ADES, H. W., AND FELDER, R. E. The acoustic area of the monkey *(Macaca mulatta). J. Neurophysiol.*, 5: 49–54, 1942.
2. AKERT, K., GRUESEN, R. A., WOOLSEY, C. N., AND MEYER, D. R. Klüver-Bucy syndrome in monkeys with neocortical ablations of the temporal lobe. *Brain*, 84: 480–498, 1961.

3. AKERT, K., WOOLSEY, C. N., DIAMOND, I. T., AND NEFF, W. D. The cortical projection area of the posterior pole of the medial geniculate body in *Macaca mulatta. Anat. Rec.*, 133: 242, 1960.

4. BAILEY, P., VON BONIN, G., GAROL, H. W., AND MCCULLOCH, W. S. Functional organization of the temporal lobe of monkey *(Macaca mulatta)* and chimpanzee *(Pan satyrus). J. Neurophysiol.*, 6: 121–128, 1943.

5. BENJAMIN, R. M., AND WELKER, W. I. Somatic receiving areas of cerebral cortex of squirrel monkey *(Saimiri sciureus). J. Neurophysiol.*, 20: 286–299, 1957.

6. BRUGGE, J. F. Auditory cortical areas in primates. This volume, Chapter 3.

7. BRUGGE, J. F., AND MERZENICH, M. M. Representation of frequency in auditory cortex in the macaque monkey. In: *Physiology of the Auditory System*, edited by M. B. SACHS, Baltimore: National Educational Consultants, 1972, pp. 283–287.

8. BUCY, P. C., AND KLÜVER, H. Anatomic changes secondary to temporal lobectomy. *Arch. Neurol. Psychiat.*, 44: 1142–1146, 1940.

9. BUCY, P. C., AND KLÜVER, H. An anatomical investigation of the temporal lobe of the monkey. *J. Comp. Neurol.*, 103: 151–252, 1955.

10. BURTON, H., AND ROBINSON, C. J. Multiple somatic sensory representation within the second somatic sensory area and adjacent cortical regions in the parietal cortex of cynomolgos monkeys. In: *Cortical Sensory Organization.* Vol. 1, Chapter 4, edited by C. N. WOOLSEY, Clifton, New Jersey: Humana Press, 1981.

11. DAVIS, H. Psychophysiology of hearing and deafness. In: *Handbook of Experimental Psychology*, edited by S. S. STEVENS, New York: Wiley, 1951, pp. 1116–1142.

12. DIAMOND, I. T., CHOW, K. L., AND NEFF, W. D. Degeneration of caudal medial geniculate body following cortical lesion ventral to auditory area II in cat. *J. Comp. Neurol.*, 109: 349–362, 1958.

13. DOWNMAN, C. B. B., WOOLSEY, C. N., AND LENDE, R. A. Auditory areas I, II and Ep: Cochlear representation, afferent paths and interconnections. *Bull. Johns Hopk. Hosp.*, 106: 127–142, 1960.

14. EARLS, F. J., AND HIRSCH, J. E. Unpublished work.

15. FRIEDMAN, D. P. Body topography in the second somatic sensory area of the monkey. In: *Cortical Sensory Organization*, Vol. 1, Chapter 5, edited by C. N. WOOLSEY, Clifton, New Jersey: Humana Press, 1981.

16. HAIGHT, J. P. The general organization of somatotopic projections to S II cerebral neocortex in the cat. *Brain Res.*, 44: 483–502, 1972.

17. HIND, J. E., BENJAMIN, R. M., AND WOOLSEY, C. N. The auditory cortex of squirrel monkey *(Saimiri sciureus). Fed. Proc.*, 17: 71, 1958.

18. IMIG, T. J., AND REALE, R. Patterns of cortico-cortical connections related to tonotopic maps in cat auditory cortex. *J. Comp. Neurol.*, 192: 293–332, 1980.

19. IMIG, T. J., RUGGERO, M. A., KITZES, L. M., JAVEL, E., AND BRUGGE, J. F. Organization of auditory cortex in the owl monkey *(Aotus trivirgatus). J. Comp. Neurol.*, 171: 111–128, 1977.

20. KENNEDY, T. T. K. *An Electrophysiological Study of the Auditory Projection Areas of the Cortex in Monkey (Macaca Mulatta).* Thesis. The University of Chicago, 1955.

21. KNIGHT, P. L. Representation of the cochlea within the anterior field (AAF) of the cat. *Brain Res.,* 130: 447–467, 1977.

22. LICKLIDER, J. C. R., AND KRYTER, K. D. Frequency-localization in the auditory cortex of the monkey. *Fed Proc.,* 1: 51, 1942.

23. McCULLOCH, W. S., GAROL, H. W., BAILEY, P. AND VON BONIN, G. The functional organization of the temporal lobe. *Anat. Rec.,* 82: 430–431, 1942.

24. MERZENICH, M. M., AND BRUGGE, J. F. Representation of the cochlear partition on the superior temporal plane of the macaque monkey. *Brain Res.,* 50: 275–296, 1973.

25. NEFF, W. D. Neural mechanisms of auditory discrimination. In: *Sensory Communication,* edited by ROSENBLITH, W. A., Cambridge, MA and New York, NY: MIT Press and Wiley, 1961, pp. 259–278.

26. NEFF, W. D., AND DIAMOND, I. T. The neural basis of auditory discrimination. In: *Biological and Biochemical Bases of Behavior,* edited by H. F. HARLOW AND C. N. WOOLSEY, Madison, WI, University of Wisconsin Press, 1958, pp. 101–126.

27. PINTO-HAMUY, T., BROMILEY, R. B., AND WOOLSEY, C. N. Somatic afferent areas I and II of dog's cerebral cortex. *J. Neurophysiol.,* 19:485–499, 1956.

28. PRIBRAM, K. H., ROSNER, B. S., AND ROSENBLITH, W. A. Electrical responses to acoustic clicks in monkey: extent of neocortex activated. *J. Neurophysiol.,* 17: 336–344, 1954.

29. REALE, R. A., AND IMIG, T. J. Tonotopic organization in auditory cortex of cat. *J. Comp. Neurol.,* 192: 265–291, 1980.

30. ROSE, J. E. The cellular structure of the auditory region of the cat. *J. Comp. Neurol.,* 91: 409–439, 1949.

31. TALBOT, S. A. Adapting a commercial cathode ray for direct-coupled, single-sweep applications. *Rev. Scient. Instr.,* 12: 100–101, 1941.

32. WALZL, E. M. Representation of the cochlea in the cerebral cortex. *Laryngoscope,* 57: 778–787, 1948.

33. WALZL, E. M., AND WOOLSEY, C. N. Cortical auditory areas of the monkey as determined by electrical excitation of nerve fibers in the osseous spiral lamina and by click stimulation. *Fed. Proc.,* 2: 52, 1943.

34. WHITZEL, B. L., PETRUCELLI, L. M., AND WERNER, G. Symmetry and connectivity in the map of the body surface in somatosensory area II of primates. *J. Neurophysiol.,* 32: 170–183, 1969.

35. WOOLSEY, C. N. "Second" somatic receiving areas in the cerebral cortex of cat, dog and monkey. *Fed. Proc.,* 2: 55–56, 1943.

36. WOOLSEY, C. N. Additional observations on a "second" somatic receiving area in the cerebral cortex of the monkey. *Fed. Proc.,* 3:53, 1944.

37. WOOLSEY, C. N. Patterns of sensory representation in the cerebral cortex. *Fed. Proc.,* 6: 437–441, 1947.

38. WOOLSEY, C. N. Organization of cortical auditory system: A review and a synthesis. In: *Neural Mechanisms of the Auditory and Vestibular Systems,* edited by G. L. RASMUSSEN AND W. F. WINDLE. Springfield, IL: C. C Thomas, 1960, pp. 165–180.

39. WOOLSEY, C. N. Organization of cortical auditory system. In: *Sensory Communication,* edited by W. A. ROSENBLITH. Cambridge, MA and New York, MIT Press and Wiley, 1961, pp. 235–257.

40. WOOLSEY, C. N. Dedication of the workshop to the memory of Edward McColgan Walzl. In *Physiology of the Auditory System,* edited by M. B. SACHS, Baltimore, MD: National Education Consultants, 1972, pp. 5–20.

41. WOOLSEY, C. N. Tonotopic organization of the auditory cortex. In: *Physiology of the Auditory System,* edited by M. B. SACHS, Baltimore, MD: National Education Consultants, 1972, pp. 271–282.

42. WOOLSEY, C. N., AND FAIRMAN, D. Contralateral, ipsilateral and bilateral representation of cutaneous receptors in somatic areas I and II of the cerebral cortex of pig, sheep, and other mammals. *Surgery,* 19: 684–702, 1946.

43. WOOLSEY, C. N., AND WALZL, E. M. Topical projection of nerve fibers from local regions of the cochlea to the cerebral cortex of the cat. *Amer. J. Physiol.,* 133: 498, 1941.

44. WOOLSEY, C. N., AND WALZL, E. M. Topical projection of nerve fibers from local regions of the cochlea to cerebral cortex of the cat. *Bull. Johns Hopkins Hosp.,* 71: 315–344, 1942.

45. WOOLSEY, C. N., AND WALZL, E. M. Topical projection of the cochlea to the cerebral cortex of the monkey. *Amer. J. Med. Sci.,* 207: 685–686, 1944.

Index